*Teaching the
Gifted and Talented
in the
Science
Classroom*

## The Authors

William D. Romey is Professor of Geology/Geography at St. Lawrence University, Canton, New York. Dr. Romey is the author of *Teaching Language Arts Creatively, Concrete Is Not Always Hard,* and *Language Arts and the Gifted Child.*

Mary L. Hibert is the Director of Science, Needham Public Schools, Massachusetts. Ms. Hibert has been a consultant in science for the MITS program, a collaboration of seven museums in the Boston area.

## The Advisory Panel

George T. Ladd, Professor of Education, Boston College

Harold L. Levitan, biology teacher, Foxborough High School, Massachusetts

Markley B. Morrill, science teacher, Willard Intermediate School, Santa Ana, California

## The Series Editor

NEA's *Teaching the Gifted and Talented in the Content Areas* Series Editor is Frederick B. Tuttle, Jr. Dr. Tuttle is Assistant Superintendent, Needham Public Schools, Massachusetts. He is also the author of *Composition: A Media Approach, Technical and Scientific Writing* (with Sarah H. Collins), *Gifted and Talent Students,* and *How to Prepare Students for Writing Tests;* and the editor of *Fine Arts in the Curriculum,* all published by NEA.

# Teaching the Gifted and Talented in the Science Classroom

Second Edition

by
William D. Romey
with Mary L. Hibert

NEA's Teaching the Gifted and Talented
in the Content Areas series
Editor: Frederick B. Tuttle, Jr.

National Education Association
Washington, D.C.

Copyright © 1988, 1980
National Education Association of the United States

Printing History:
   First Printing:        August 1980
   Second Printing:    July 1981
   SECOND EDITION:  January 1988

**Note**

The opinions expressed in this publication should not be construed as representing the policy or position of the National Education Association. Materials published as part of this NEA multimedia program are intended to be discussion documents for teachers who are concerned with specialized interests of the profession.

Library of Congress Cataloging-in-Publication Data

Romey, William D., 1930–
   Teaching the gifted and talented in the science classroom / by William D. Romey, with Mary L. Hibert. — Sec. ed.
   p. cm. — (NEA's teaching the gifted and talented in the content areas series)
   Bibliography: p.
   ISBN 0-8106-0748-4
   1. Science—Study and teaching. 2. Gifted children—Education. 3. Talented students—Education. I. Hibert, Mary L. II. Title. III. Series.
Q181.R66 1988
507'.1—dc19          87-16114
                    CIP

# CONTENTS

Introduction to the Second Edition ................................................. 7

Preface .................................................................................. 8

On Being Gifted and Talented ................................................... 10
    Excellence ......................................................................... 10
    Testing, "Normal" Distributions, and Giftedness ...................... 11
    Motivation ........................................................................ 13
    A Model for the Growth of Intellect, Skills, and Giftedness ........... 14

Characteristics of the Gifted ..................................................... 19
    A "Catalog" of Types of Giftedness ...................................... 20
    What Makes a Practicing Scientist Gifted? .............................. 21
    Giftedness as a Burden ....................................................... 24

Teaching Science ..................................................................... 26
    Progress in Science Through the School Years ......................... 26
    Scientific Literacy and the Gifted .......................................... 28
    Science as "Serious Business" or "Play"? .............................. 29
    Competition and Giftedness ................................................. 29
    Using Programs for the Gifted as Routes for Change ................ 31
    Giftedness and Advance Placement in Science ........................ 32

The Learning Environment ....................................................... 33
    The Group and the Individual .............................................. 33
    An Open-Access Educational Environment in Science ............... 34
    Problem-Centeredness and Student-Centeredness ..................... 37

Science Activities for the Gifted ................................................. 41
    Exercises in Perception, Awareness, and Observation ................ 41
    Fantasy Exercises .............................................................. 44
    Values and Perception ........................................................ 45
    Making the Strange Familiar and the Familiar Strange ............... 46
    Hard Eyes—Soft Eyes ........................................................ 47

    Science and Interconnectedness ............................................... 48
    Exercises in Applications of Skills ............................................ 49
    Creativity Exercises and Games ................................................ 51
    Illustrating Ideas Through Other Media .................................... 53
    TV—If You Can't Fight 'em, Join 'em ...................................... 54
    Mental Maps ............................................................................. 55
    Answers and Truth ................................................................... 56
    Ambiguity and Uncertainty ..................................................... 57
    Stepping Outside the School Environment ............................... 58

What About the Basics? ................................................................. 60

Bibliography ................................................................................... 61

Selected Resources ......................................................................... 64

# INTRODUCTION TO THE SECOND EDITION

In modifying the first edition of this text, Mary Hibert has based her additions on a specific philosophy of science education:

> The purpose of science education is to develop scientifically, technologically, and culturally literate students for active participation in our modern society. To achieve this goal, the science curriculum should provide young people with learning situations that give them first-hand experiences with natural phenomena and permit them to gain more information about the world. Logical and creative thinking skills and a scientific attitude should be a major part of a thinking process that incorporates inquiry, problem solving, decision making, analysis of data, and drawing of conclusions. These skills should enable students to acquire, process, and utilize information in the context of thinking critically, making decisions, and forming ethical judgments. The science curriculum should be dynamic, open-ended, and constantly challenging, providing many opportunities for hands-on activities.

This philosophy echoes the ideas and activities presented by Bill Romey in the first edition of *Teaching the Gifted and Talented in the Science Classroom*. Both writers are vitally concerned with the creative, exploratory, risk-taking nature of science rather than with the acquisition of facts about science. Not only does Ms. Hibert illustrate her philosophy in her additions to the text, but also she practices it in her position as Director of Science in the Needham (Massachusetts) Public Schools. The entire elementary science curriculum in Needham revolves around activities and hands-on units developed through the Needham Science Center. Students learn through experimentation: observing, analyzing, discussing, concluding, and revising without the use of textbooks.

To enable students to learn through this approach, the teacher and the school system must create an appropriate environment. Such an environment encourages students to try ideas even if they might fail. Indeed, one may argue that we learn more from errors than from successes, especially if we accept errors as valuable learning experiences. As Bill Romey states, we should "look closely and nonjudgmentally at 'mistakes' to see what hidden messages may lie within them...." To encourage creative thinking, then, we have to allow students to fail. This attitude is particularly evident in the activities presented throughout this text.

*—Frederick B. Tuttle, Jr.*
Series Editor

# PREFACE

As I began my work on this book, the title *Teaching the Gifted and Talented in the Science Classroom* worried me a little. My approach here does not lend itself entirely to the word "teaching" as conventionally defined. As I started to type "T-e-a-c-h-i-n-g" the word "R-e-a-c-h-i-n-g" appeared. Was this a simple slip of the finger? After all, "T" and "R" are next to each other on the keyboard, and both are struck with the same finger. Or was this a "directed" mistake, a purposeful redirection of my finger by some preconscious autopilot? Whatever the origin of this "mistake," it is an example of what Carl Jung called "synchronicity." (29)* In simplest form, synchronicity is a coincidence in time and space of objects, things, or ideas in which a person finds meaning. Once we are open to such happenings, we find many instances in our lives—small coincidences of all sorts. Some people think of these coincidences in terms of "destiny" or seek a cause behind them. Jung, on the other hand, thought of them as "acausal." That is, their meaning is one we assign to them rather than an objective, inherent meaning that is a result of their cause-effect relationship.

But what does this have to do with teaching science to gifted and talented students? For one, it challenges the cherished scientific notion of "causality." Most traditional science instruction is based on the notion that the function of science is to seek causes. Synchronicity stresses questions that ask "How?" rather than questions that ask "Why?" "How" questions deal with relationships and processes and do not demand justification and explanation. They tend to accept the world as given.

Progoff (29) suggests there may be a high incidence of synchronistic events in the lives of people classified as "creative." My own preliminary tests among practicing creative scientists tend to confirm this idea. This carries important implications for education of the gifted. Progoff notes the need to create an environment of mutual trust, caring, and support in order to deepen the quality of group and individual experience. Only when individuals have no fear of "exposure" and ridicule can they become fully sensitive to synchronistic experiences and willing to share them.

---

*Numbers in parentheses appearing in the text refer to the Bibliography beginning on page 61.

### Activity: Isn't It a Coincidence?

Ask students to keep a log of the coincidences that occur in their lives over a week. Keep a log of your own coincidences to share with students. Once you show you are willing to share some personal coincidences, your students will feel free to share theirs.

Keep a "coincidence box" in the corner of the classroom. Students can jot down reports of their coincidences on a piece of paper and drop these anonymous notes into the box. Set aside time to read the contents of the box aloud to the group.

Whenever a cause-effect situation presents itself, ask students to analyze whether one event really caused the other. Perhaps the events just happened to be associated in time or space.

Find some "coincidences" in biology, chemistry, physics, and the earth sciences. How are these related?

Several points relating to gifted students emerge from the implications of my typing "error." Gifted students often have a different view of the world. This is why we work with them in ways that are different from what goes on in ordinary classroom situations. A first point is that an idea or solution may be the result of a "mistake" rather than a result of "correct" work. A second point is that ordinary diligence and attention to detail may actually deprive one of a new idea or lead one away from a promising direction. Third, one needs to be fully attentive to what one *is* doing rather than what one *should* be doing. Fourth, trust preconscious instincts; the unexpected things that emerge may be useful. Fifth, let behavior follow natural channels of energy flow, the channels the body chooses to follow rather than the ones the "rational mind" tries to impose. Sixth, look closely and nonjudgmentally at "mistakes" to see what hidden messages may lie within them; sometimes correcting a mistake leaves you with an ordinary product rather than something more creative and more to the point. Seventh, move ahead on ideas, get them down, play with them, and use them as a springboard for progress and growth. Eighth, remember that small happenings can lead to bigger and often more useful ideas. Teachers who follow this approach will be helping gifted students move rapidly and surely toward realizing their potential.

*—William D. Romey*
*East Orleans, Massachusetts*

# ON BEING GIFTED AND TALENTED

According to Goethe, "If you treat an individual as he is, he will stay that way, but if you treat him as if he were what he could be, he will become what he could be." Every person is probably gifted or talented in some way. It is the primary job of an enlightened system of education to help individuals discover and develop their areas of latent giftedness. (50) If this objective is not met, both the individuals and the society are denied access to an invaluable resource. Sad to say, educational systems have, to date, been notoriously inattentive to this responsibility.

## EXCELLENCE

Bertrand Russell (43) laments the American conception of democracy in which all people are regarded as alike and people who demonstrate excellence in some areas are regarded as setting themselves up as superior to others. In this system practically the only people allowed to be real "aristocrats," especially in school situations, are outstanding competitive athletes. Of course, there are a few other areas where people compete for prizes. In general, however, the prevailing desire is to "blend in," to be like everybody else. People tend to hide special talents rather than display them. As a result, talents often atrophy from lack of use. This phenomenon has been especially well documented in the case of girls and women who downplay their intellectual power. Although the situation is changing now, our sexist society to some degree still specifies a largely submissive and passive role for females. Consequently, many persons overlook or even deny the exceptional abilities of women, especially in mathematics and science.

### Activity: Looking for Hidden Excellence

Put an "excellence box" in the back of your room. Invite your students to deposit anonymous notes indicating what they feel they do well. Invite them to also deposit notes indicating what a classmate does well but doesn't advertise. Check the box periodically and let the class know that there is someone in the group who can do _____(whatever it is) well.

Assignment to class: Do something you do well during some class activity. Don't tell the others you are going to do it; just do it and enjoy doing it well. Afterwards drop an anonymous note in the "excellence box" to let the class know what thing you did excellently but without fanfare.

# TESTING, "NORMAL" DISTRIBUTIONS, AND GIFTEDNESS

Ironically, within this democratic ethic of students, where no one is supposed to be better than anyone else, has grown the devastating "scientific" notion of "normal distribution." Beginning with standardized testing and moving down into the evaluation systems of ordinary classrooms, we have come to believe that 10 or 15 percent of the students must be labeled "best" and an equal percent must be labeled "worse." All the rest occupy a never-never-land of "better-than-average/average/less-than-average." Worse still, performance in areas as diverse as music, athletics, literature, language, arts, communication, science, and social studies all get lumped together to form a meaningless general average for each student, and these averages are used to establish equally meaningless group norms. Many of the tests that measure performance examine people for a narrow range of skills and neglect other skill areas that may be of equal or greater significance both to the individual and to society. These testing systems traditionally penalize people for what they don't know or don't do well rather than seek to help people discover what they do know and can do well.

This testing system ultimately labels 25 percent of the population as good or excellent; 75 percent of the population are left to regard themselves as only average, below average, or poor. Fortunately this approach to testing has come under serious challenge, at least for the big standardized tests such as the SAT, LSAT, GRE, MCAT, and so on.

Fiske (11) has summarized the New York legislation that requires test preparers to open the curtain of secrecy that has shrouded them for years to explain exactly what their statistics mean and disclose how great are the errors that so strongly affect the lives of virtually every student in the United States. At the present time, after taking the SAT exam, students may ask for a copy of their answer sheets and a copy of the correct answers. Fiske reports that in October 1979 a Federal district court in California ruled against the use of standardized intelligence tests as criteria for labeling children as mentally retarded. An equally compelling argument could be made against using such tests as a basis for determining who is "gifted." (50)

Some people clearly are gifted; they do some things better and more easily than other people. Of course, no one does everything better than everyone else. Some people, however, manage to do many things very well. The Olympic athletes who compete in the decathlon and pentathlon show exceptional skills in several sports. Yet their scores in any one of those sports generally lag behind those of front-runners in the individual sports. Even being a jack-of-all-trades is a "special talent."

Giftedness cannot be defined in a blanket way for the population at large. IQ tests, standardized tests, and statistical treatments may be convenient for sorting people into groups and avoiding the appearance of chaos and infinite individuality, but they are false indicators. They accomplish more harm than good, both for the individual (especially if he or she is at the lower end of the distribution) and ultimately for society. In general, the only people they really benefit are those in personnel departments and school admissions offices.

### *Activity: Normal Distribution*

Have students gather a large number of *similar* objects (leaves, pebbles, books, or something that is part of your regular lesson). Ask groups of students to mark each object in some way so that each has its own special mark. (Number them, for example.) Then ask each group of students to "grade" the objects A, B, C, D, or F, so that 10 percent get As, 15 percent Bs, 50 percent Cs, 15 percent Ds, and 10 percent Fs, with A being "the best" and F being "the worst." When all groups have had a chance to do the grading, compare the grades for specific items. Did a particular item receive the same grade in each group? Discuss how value judgments affected the assigned grade.

Ask students to gather a large number of *different* objects. Mark these objects and go through the same procedure. What happens when one tries to compare and assign grades to very different things?

This exercise demonstrates the fallacy of normal distribution when applied to individual items. It also shows the futility of trying to compare "apples and oranges" (that is, unlike items) on the same grade scale.

Teachers must assume that each student who comes their way, no matter what age or grade level, is gifted or talented in some way. Whether we are teachers in a local Head Start program or in the most elite of graduate schools, we must not ask, "Is this student gifted?" but rather, "In what areas is this student gifted, and how can she or he use the areas of my 'course' to express that giftedness, increase the vitality of the classroom environment, and work toward personal growth and maximization of individual potential?"

Teachers who espouse this view of giftedness can boost students' self-concepts. Rosenthal and Jacobson (41) have shown that students tend to behave according to the expectations of the people around them. When teachers treat students as "gifted," the youngsters tend to behave like gifted students. It is equally clear that students treated as stupid will behave that way.

A student may appear to be gifted mainly because she or he has already had practice in that specific area. Another student who initially seems less apt may, after becoming familiar with the task or area, perform better than the first student. It is a mistake to reward the first and take no notice of the latter, because without encouragement the late bloomer may never exert the energies necessary to develop that particular potential. This brings us to the matter of motivation.

## MOTIVATION

One of the basic criticisms levied against standardized testing is that it takes no account of motivation. A person may have a "natural" facility in a certain area but may not be interested in developing that skill. A student should not be pushed to develop an ability on the basis of test scores or other such criteria; this may not help the learner and could even do harm.

On the other hand, some students are highly motivated to excel in an area even though their test scores are low. When these students do well we mistakenly label them as "overachievers." Telling a student that he or she really hasn't the ability to do so well is one of the most gratuitous insults perpetrated upon students in the name of standardized testing.

The development of any part of a person's potential depends on two basic factors: aptitude and motivation. I believe that motivation is by far the more important. A third factor, the environment, is perhaps the equal of the other two. The environment can help the student by providing freedom to develop in a given direction and by offering appropriate encouragement, assistance, and necessary resources. However, one must be watchful for conditions in the environment that push children .further in a direction that they have temporarily or permanantly abandoned, because failure to acknowledge changes in children's motivation can have a serious negative impact on their growth.

### *Activity Motivation*

Ask students to keep track of how long it takes them to do 10 different things they really don't want to do. For each task, ask them to write a few sentences describing the activity. Have them keep the same kind of record for 10 activities they really want to do.

Randomly select some of the papers (they can be submitted anonymously) and discuss with the group some of the differences found. Are there any instances where one student really wanted to do an activity that another student was totally unmotivated to do? Note the time differences reported.

Consider several of the tasks people really didn't like to do and discuss ways in which these activities could be made interesting and desirable. Then consider some tasks people enjoyed and discuss how these could be made undesirable.

On the chalkboard write a "Question of the Week"—such as "How many mirrors in a kaleidoscope?" "What is the difference between a toad and a frog?" "Why aren't spiders caught in their own webs?" Ask students to find the answer to the question, and at the end of the week discuss it. If possible, carry out an activity to determine the answer.

## A MODEL FOR THE GROWTH OF INTELLECT, SKILLS, AND GIFTEDNESS

Each person brings to a learning environment certain areas of knowledge and skills developed in previous experiences at school, home, play, and elsewhere. Skill in a particular area does not always signal a high level of interest in that area. Another dimension might be termed level of aptitude. These three factors are charted for a typical student named Jane (see Figure 1). Teachers using "standard" approaches to determine giftedness would probably channel Jane toward work in music and science. Is this the best course for Jane? She has spent a good deal of time in music, has developed her skills to a moderate degree, but has low interest. Her prior experience mainly consists of piano lessons she was forced to take. But music is a whole universe of related activities. Perhaps Jane loves to sing or might do well with the guitar or another instrument. So it is not really valid to say that she is not interested in music at this time, based only on her limited experience with piano. A generalization of this sort may cause us to ignore an entire field in which Jane could excel.

Now compare the "science" chartings in Figure 1. Jane shows high "aptitude" in science but has spent little time in that area and has developed few skills. Perhaps the low interest in science is related to her lack of experience.

Aptitude tests can probably be trusted when they give a high score. A low score, on the other hand, probably should be viewed with suspicion. Telling a student that he or she has "low aptitude" in any area probably is enough to destroy any budding interest in that area. The best route to encourage growth is to stress the positive factors and leave all other factors open. Any attempt to force students into areas where they *appear* to have talent, in the absence of demonstrated interest, is likely to destroy rather than boost motivation.

Figure 1 shows then that for each student a spectrum of factors applies: level of development, level of interest, aptitude (as indicated by testing or as "sensed" by parents, teachers, or others), and amount of energy previously spent in the area. Each subject area considered in such an analysis is open to great expansion. A person who shows up low on "science" may show up high on "insects" or "dinosaurs" or

# Figure 1. Factors for Assessing Giftedness

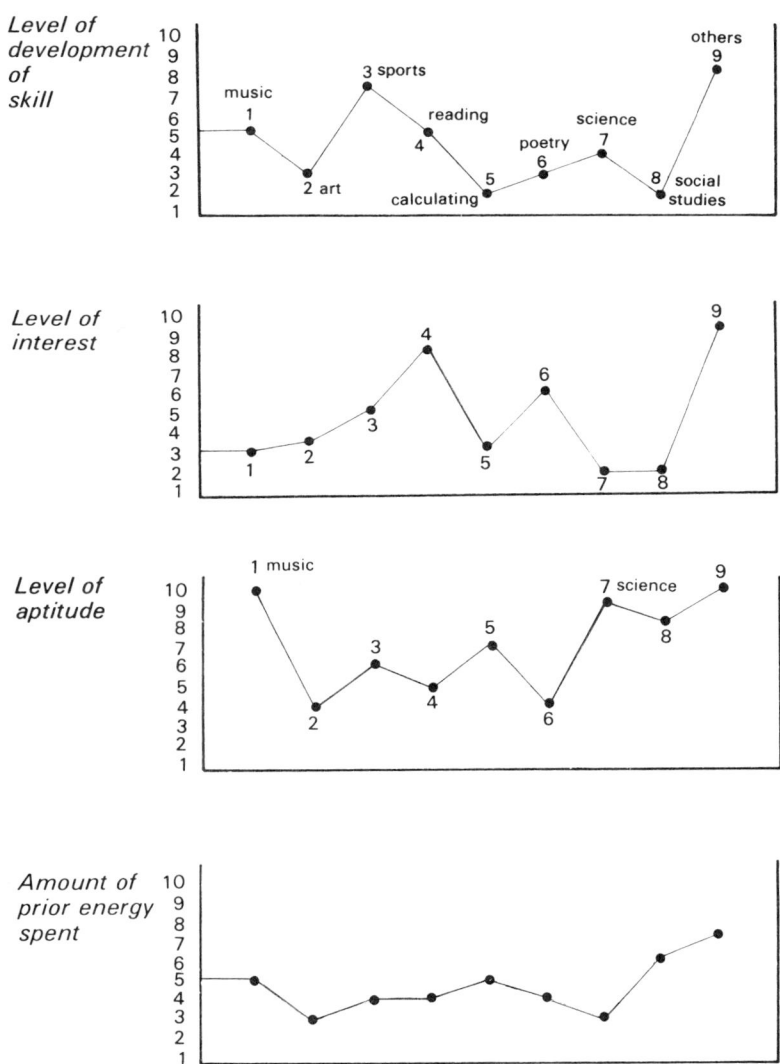

15

"electricity." In my own case, I hated "science" as an elementary student, high school student, and undergraduate. Yet my graduate training eventually led me to become a professional scientist. A student low in "reading" may be high in "dramatics" or "science" or some other area in which reading plays a significant part. The analysis of even one student is highly complex, and the possibilities of finding an area of giftedness are enormous.

### Activity: Where Is the Gift?

Have each student prepare a set of four charts similar to those in Figure 1. The students can use these charts as a basis for an exercise in self-awareness and as a way to find out more about each other's interests and backgrounds.

It probably is possible to keep track of several different factors that relate to the appearance of "giftedness" in learners. But the common system of recordkeeping, based on letter-type grades, cannot provide the information needed for a full analysis. If anything, the present grading system can mislead more than it can help. However, a teacher intent on completing a full analysis would have to spend so much time keeping track of the various factors that little time would be left to work with students. Better to leave this kind of research to the psychometricians (who are "gifted" in the manipulation of numbers and curves) and devote teaching time to the positive facilitation of learning, based on the assumption that each student is gifted in more areas than we can imagine. The discovery of a new area of interest and talent can be a source of excitement and joy for student, teacher, parents, and classmates. Each such discovery merits its own special celebration.

### Activity: Party!

Organize a brief celebration in your class when a student or students discover a new interest. In a larger "unit party" students might come dressed to represent their new interests or dramatize, draw, or otherwise exhibit their interests. Encourage student participation but don't require it. Participate yourself; this may help timid students become willing to celebrate more openly.

## Some Myths to Explore

There are two sorts of myths to expose here: some that deal with the nature of giftedness in science and some that deal with the nature of science itself. One common myth is that boys have more aptitude for science than do girls. The evidence from many studies supports this view, but when the assumptions behind the research are analyzed, one comes to doubt the validity of those studies. Such studies are based on standardized tests that do not take into account the sex-related conditioning that leads young boys and girls to behave differently from one another. Girls' toys, clothing, and games all conspire to orient them away from the kinds of activities traditionally associated with science. Thus, the gender differences in science aptitude and skills are likely to be related to upbringing rather than to latent potential. When we treat girls as if they are less capable in science than boys are, we ensure that they will perform less well. The prophecy fulfills itself.

Another myth is that boys do better than girls in mathematics, mechanical skills, and three-dimensional visualization. I recently worked with a college woman in a geology program. This woman talked about her lack of aptitude in these areas. Nevertheless, a letter from a professional geologist who supervised the work of this woman praised her ability to see things in three dimensions, a talent which she resolutely denied having "since she was a girl." The best place to explode these sexist myths is in the home, during the first years of life. Elementary school teachers also have a good chance to catch these attitudes early. Each of us, however, at whatever level, is obligated to recognize and deal with these attitudes if we are serious about recognizing latent giftedness and helping students develop their potential.

A myth related to the nature of science is implicit in the previous example. This myth holds that scientific questions and their answers are value-free: if you ask a question "scientifically," the answer can be "believed" until proved wrong. There is a basic flaw in this kind of reasoning. We must look at the assumptions that precede the asking of a scientific question before we look at the results. The range of answers possible in any scientific study is limited by the nature of the initial question. Major scientific advances are the result of asking supposedly "old" questions in new ways that expand the range of possible answers.

### *Activity: Seeking the Hidden Assumptions*

Ask students to choose any major question presented in their textbook. Have them determine what assumptions led to the framing of that question. Then suggest that they start from a different set of assumptions. What would the question look like then?

(Examples: Any question that asks, "How does X affect Y?" assumes causality, which may not exist. "How do radioactive atoms decay?" assumes that "atoms" exist and that they "decay" when radioactive. The words themselves are metaphoric labels, which, if changed, present the question in a new light. "Where do the worms go in the winter?" implies that they do go somewhere. It also assumes the validity of the classification "worm." And what about "winter"? It's a valid concept in some northern areas, but how about the equatorial regions?)

## The Problem with a Narrow View of Giftedness

One problem with a narrow definition of giftedness is that it makes certain occupations seem more worthy than others. The common emphasis on academic giftedness is reflected in the status accorded to certain professions: medicine, law, engineering, and, to some extent, science. As a result, schools tend to neglect students who have special talents that lead them in other, less favored directions. This is particularly true in the United States; in some parts of Europe each profession has its own standards and traditions but does not consider itself superior to other professions. (43) What would the world be like if everyone managed to get into that "highest" profession? It would be like an orchestra in which everyone was the concertmaster. As Bertrand Russell says, "There does not seem to be an adequate understanding of the fact that society should be a pattern or an organism, in which different organs play different parts. Imagine the eye and the ear quarrelling as to whether it is better to see or to hear, and deciding that each would do neither since neither could do both." (43) Society cannot function effectively without gifted practitioners in all areas; each person has a place.

### Activity: The Non-Co-op

Assign the class a complex team-oriented job such as putting out a science journal or a newspaper. Tell students that you are passing out team assignments but that no one is to tell anyone else what assignment he or she received. Then hand everyone a folded slip with the word "editor-in-chief" written inside. Tell the group they have one or two class periods to produce a first draft. Afterward discuss the process and implications.

### Activity: The Body

Have students perform a drama called "The Body." Assign each person the role of some organ or cell type. Ask them to research in advance exactly what their role in the body is. Then have them act out "the body at work." They might try such plots as "circulation," "digestion," or "injury and recovery." Each person plays a cooperative role that benefits the body as a whole.

Now ask half the class to perform a single vital part (for example, the heart). Do not replace their previous functions. Then ask three-quarters of the class to perform a single vital function, with additional functions going undone.

Discuss the nature of differentiation as an essential quality of any complex organism, with all parts being equally important in the long run.

Try other plays using other metaphors in which all parts are necessary to the functioning of the whole. These exercises offer important scientific content as well as emotional and physical activity.

# CHARACTERISTICS OF THE GIFTED

Two other myths about science relate to qualities teachers might think signify giftedness in science. The first myth is that science is an orderly discipline that progresses logically from one step to another. Many scientists claim this is true and belong to a group T.S. Kuhn (21) has labeled "normal scientists." But not all scientists share this view. Kuhn points out that science makes its major advances by leaps and jumps rather than in a simple, linear fashion. Giftedness in science often reflects a willingness to break the rules rather than follow them. Of course, it takes orderly people to do much of the work of science, but we'd be nowhere without those gifted rule-breakers. Paul Feyerabend (3) is another philosopher of science who shares Kuhn's view, taking it to further extremes. In his view the advance of science depends not only on rational argument, but also on propaganda, rhetoric, and subterfuge. Feyerabend (10) believes in a strong methodological pluralism in science. Thus, giftedness in science might go hand in hand with such qualities as craziness, playfulness, argumentativeness, explorativeness, and resistance to authority. A strong creative bent is as important in science as in theater or art. The environment in which gifted students work on their science should encourage the expression of these qualities as well as the demonstration of rational analysis, systematic observation, and the other virtues traditionally associated with science.

The second myth concerns neatness in science. In advertisements, in journals, and in television we are confronted by the image of the scientist in a clean white lab coat. Some scientists do dress this way. But it is more realistic to show a cluttered laboratory populated by people in highly informal dress, who joke around a good deal, and whose lab notes might be written on odd scraps of paper. The standard image of the neat scientist is not necessarily a wrong one; it is merely incomplete. The scientist must be seen as a complete person if we are to recognize potential for giftedness in messy secondgraders or untidy teenagers. We must create environments—often messy and disorderly—in which this giftedness can flourish.

### Activity: The Scientist as a Person

Ask students to describe "the typical scientist." Does a stereotype emerge? What is it like? List these characteristics and keep them. Now ask students to gather from newspapers, magazines, films, TV, novels, and other sources information about scientists as people and about their laboratories and their work. A book like Watson's *The Double Helix* (54) can be useful. Does this new information change students' assumptions about scientists as people?

# A "CATALOG" OF TYPES OF GIFTEDNESS

1. *The all-around genius.* To many people, the student who fits this stereotype is highly precocious, temperamental, and difficult to get along with. In truth few students fit this extreme, but in those who do, psychological problems probably derive from the way they are treated by others and not from their giftedness as such. In dealing with these students, the best course is to encourage their growth and keep out of their way while they pursue their own strong interests. Teachers can help them find balance by introducing into their environment elements and people who can help broaden what might otherwise turn out to be narrow "genius."

2. *The person gifted in science but not in other areas.* Such students play science games all the time, pay little attention to other schoolwork, and are continually told by others that they are good in science. They pose a problem in that they don't get put into a special class for the generally gifted and yet clearly need special resources and expertise to guide and encourage their talent in science. Often students who show exceptional precocity in a single area may use their recognized abilities in this area to get out of other things. Parents and teachers may acquiesce and stop providing challenges and opportunities that may bring out talents in other areas. The students continue to do what they already know they can do better than others and become less willing to risk trying new activities in which they may appear less advanced or less special. Sometimes it helps to broaden the definition of science to introduce the very risks the student has tried to avoid. Or introduce a "non-science" activity as part of a regular science class; gifted students may find themselves doing well before they realize that they have made the jump to a new area.

3. *The person gifted in other areas but apparently not gifted in science.* A student's apparent lack of giftedness in science may be due to lack of aptitude, lack of motivation, lack of interest, or prior bad experiences. Usually, however, the problem is one of semantics. The word "science" may induce "science anxiety," a state of mind similar to the well-known "math anxiety" or "reading anxiety." A way to explore for latent giftedness is to encourage other teachers to introduce

content that is in fact "science" but does not carry that name. If students can be placed in a gifted program without specified class times and without classes labeled by discipline, content and skills normally the province of science can appear in literature, environmental work, art, music, and other areas. Later, after students have become competent in several areas, they are often amazed to realize they've been doing "science" all along and actually now seem "gifted" in this area, too.

4. *The "average student."* Most students fall into this category. They are well-balanced, follow the rules, do what they are supposed to do, get average grades, and don't make trouble. They never attract much attention. They fit in and pose no special problems. Yet there is a great reservoir of hidden giftedness within this group. It is the unmet challenge of public education to find and encourage the growth of this untapped talent.

5. *The handicapped and "retarded" as "gifted."* Research indicates that many students currently classified as handicapped or retarded are actually gifted in certain areas but lack the "balance" that makes other children seem "normal." Thus special education teachers may also find the suggestions in this book helpful in reaching those special, submerged areas of giftedness in students who are deprived or "different" or thought to be retarded. In the long run, it is essential to avoid any labels that might adversely affect the child's self-concept.

### Activity: Interdisciplinary

To encourage "giftedness" in all areas, plan a number of interdisciplinary activities that focus on science topics. Ask students to do a creative writing project—such as "Life as an Amoeba"; or after experiencing the metamorphosis of a frog or a butterfly, ask students to write a poem explaining the process or to create a mobile illustrating the process.

## WHAT MAKES A PRACTICING SCIENTIST GIFTED?

Teachers who are interested in assisting students gifted and talented in science probably should have some idea of what constitutes "giftedness" among scientists. Is it possible to isolate some of these qualities from looking at scientists who have made important contributions to science? What about those scientists who are regarded by their peers as gifted even though the complex circumstances of scientific discovery have not blessed them with precedence or fame?

Giftedness manifests itself in so many different ways in the actual practice of the sciences that one might despair of ever coming up with a general list of qualities. Many of the most gifted scientists were never recognized as having any particular gift during their schooling. This suggests that attempts to discover and coach those students who are destined to become gifted future scientists may be a waste of the teacher's time, since it is so difficult to predict who these people will be. But because we are committed to helping students uncover and use their gifts, we must proceed. Abraham Maslow (25) recommended that we not shirk from a task because it is difficult. Unimportant tasks are not particularly worthy even when done well. On the other hand, important questions need to be faced, no matter how difficult they may seem or how slim the chances of getting a satisfactory answer.

The most important scientists have been risk-takers. They enter "where angels fear to tread." They are often singleminded in their pursuit of a goal, even to the point of neglecting other responsibilities. Sometimes they appear unwilling to listen to what other people say is "reason." While Einstein was developing his most important theoretical contributions, he told people not to confuse him with the facts. (6)

Gifted scientists tend to jump over steps in the reasoning process; they make intuitive leaps. Many gifted scientists of the past have admitted that their dreams have provided routes to their discoveries. (18, 52)

In spite of their singlemindedness when "on a track," many scientists have also shown considerable gifts in other areas. Einstein was a fine violinist and became an active and persuasive figure in politics. Linus Pauling earned a Nobel Peace Prize as well as the prize in chemistry.

Many gifted scientists have shown themselves to be meticulous in their recordkeeping and careful testing of ideas. Thus, it would seem that some combination of easy access to intuitive powers and an ability to attend to the often tedious task of gathering and testing information has been important to many scientists.

A well-developed sense of curiosity is another characteristic common to gifted scientists. This includes an alertness not only to the immediate world around them but also to work that other people are doing and its implications for their own work.

Gifted scientists recognize the interconnectedness of events and relationships that casual observers fail to notice. This is what discovery really is: the bringing together of elements previously thought to be separate, the new combination of ideas or events.

Although gifted scientists are somewhat different from gifted poets, artists, novelists, and other creative people, the similarities between these groups far exceed the differences in subject matter and approach. Gifted scientists, as portrayed in the history of science, in their own autobiographies, by other biographers, and in the media, approximate other "self-actualizing people" as characterized by Maslow. (14)

Kuhn's (21) analysis of how scientific ideas develop suggests that scientists can be divided into two very different groups. In the first group are the so-called "normal scientists." Much of their work is deductive, and they tend to operate within a group of theories. The great majority of working scientists belong in this category. Another smaller group includes those scientists who introduce the big new ideas that challenge and contradict the ruling ideas.

Science progresses, say Kuhn (21) and Feyerabend (10), not by following method but by breaking it. Science moves ahead by leaps rather than in a tidy, slow, progressive way. Certain times are more favorable than others for "scientific revolutions." Once a new idea becomes established and acquires the status of dogma, it takes many years of information gathering before the problems posed by the new idea show up. There must be a certain body of data in hand before discrepancies or anomalies can be seen. A discrepancy by definition implies that a big body of information already exists. No revolution can occur if there is no pre-existing idea against which one can revolt. It takes special courage to propose that a widely accepted idea no longer holds true, especially when you have only a few discrepancies in hand as a basis for challenging the prevailing idea. Kuhn suggests that these iconoclasts are very different from "normal scientists" (21), and even though most scientists combine some of each element in their personalities, they probably lean strongly toward one pole or the other. In fact, the two types may not get along very well. In looking for giftedness we must remember these two different models.

Another factor that complicates the recognition of giftedness among scientists is the extraordinary variety of human activities that falls under the heading of science. Even the "approved" disciplines of physics, chemistry, biology, and earth sciences can be broken down into many subdisciplines that require very different human qualities from their practitioners. A great difference exists between the observational field naturalist and the theoretical physicist. Yet these two occupations are lumped together as scientific. It is comparable to putting automobile mechanics and concert violin playing under the same heading. When we add psychology, economics, and the social sciences to the physical and biological sciences, the picture becomes even more muddled. Yet, the National Science Foundation supports all of these fields. What about astrology, parapsychology (24), acupuncture, and other fields? Feyerabend (3), Rogers (31), and others agree that scientists must be open to developments in an increasingly broad range of disciplines.

Because of the wide range of opinion about what constitutes giftedness among professional scientists and because of the extraordinary breadth of science itself, it is clear that teachers interested in helping students gifted in science must be open to a wide range of behaviors and attitudes. Giftedness appears in many forms and in many kinds of people. Because giftedness can manifest itself in so many different dimensions, we must be alert to recognize it and ready to encourage its development. Where we sense imbalance, we must help learners find their centers. We must provide learning environments that foster the risk-taking necessary for growth and balance. Talents that are buried serve no purpose. Those that are uncovered and used multiply. Our job is to encourage that multiplication.

### Activity: The Catalog of Gifts and Giftedness

Have students prepare the kind of catalog sent out by department stores for special sales. Their catalog will advertise various types of exceptional skills "for sale." For each skill they should prepare an advertisement that might "sell" that particular skill. In a second section of the catalog, have them prepare statements about how those skills might show up, especially in science. Prepare ads for this section, with each person trying to sell his or her skills.

Then have an auction. Some skills can be grouped in "packages"; some can be sold separately. Give each student some "currency" to be used for buying the skills. Have one student act as a auctioneer. After the auction students can discuss how they feel about what they wanted to buy, what they were able to buy, the categories of skills available, etc.

### Activity: What Makes a Scientist Gifted?

Ask students to research biographies of scientists. (You may wish to limit this assignment to a specific area the students are presently studying.) Ask students to list the *personal* characteristics of these people. Their scientific work aside, what were they like as people?

Ask the group to watch science-oriented television programs like "Nova." They should list the personal characteristics of the scientists represented.

Finally, ask students to read parts of *The Night Is Dark and I Am Far from Home*. (20) This book shows how the images of many well-known personalities are "sterilized" for student consumption so that the real, human qualities of the people are lost. Several scientists, Thoreau among them, are represented in this book.

## GIFTEDNESS AS A BURDEN

Once we identify a person as gifted, we must be careful to make this designation a help and not a burden. When the designation carries with it obligations and responsibilities the student has not asked for, we risk inhibiting rather than facilitating the giftedness. Gifted people must be encouraged to develop their gifts. However, when this encouragement takes the form of coercion, as it often does, students may find ways of demonstrating that they are *not* gifted as a way of escaping the burden. Or they may develop unhealthy resentment against the people who are coercing them, even as they go on to develop the gift. It's a matter of simple psychology: people often resist what they're told they must do. The environment must be one of invitation rather than one of expectation and coercion.

Being gifted also can seem like a burden when persons who have been "set up" as "gifted" learn that their ability isn't so extraordinary. For some, this shock comes when they are suddenly transferred out of the gifted class back to a regular class. Or perhaps through competing in science fairs or meeting people from other schools, students see firsthand that their work pleases them much less than what others have produced. Of course, it is possible that these situations might make students work harder to "prove" continuing giftedness. But it is equally possible that the students will experience a sense of failure and inadequacy they never would have felt had they not been treated as someone of whom great things were expected. Labeling students gifted and then later telling them, "Oh, you're not really gifted after all," is the worst kind of destructive and unearned put-down.

"Giftedness" is a quality to be borne lightly. Taking it seriously carries great peril. Better by far that we treat all students as if they were gifted, taking delight in excellent work they do, without the need for labels.

# TEACHING SCIENCE

## PROGRESS IN SCIENCE THROUGH THE SCHOOL YEARS

David Hawkins (17) stressed the importance of allowing young children plenty of time for "messing about" in science. Hawkins recognized the importance of play in helping children find areas of interest. Alert teachers, by providing materials with which a child can extend play, can encourage more focused activity. In response to Hawkins, projects such as the Elementary Science Study (ESS) group created a wealth of aids, kits, and pamphlets to help teachers encourage the diverse play that must precede focused activity in the very young. The best of "serious science" also includes much unfocused "messing about," a fact overlooked in science classes for older youngsters. Other federally supported projects for elementary schools, "Science—A Process Approach" (SAPA) for one, provided some opportunity for play but attempted to focus the child onto a narrower track. SAPA appears now in few elementary schools, and usually kits remain unused in the back of the classroom as window-dressing for parents. In 1986 the federal government funded a number of projects to develop current elementary science units using a process approach.

Beginning in the upper elementary years and into middle school, students receive progressively more focused science content, much of it repeated year after year, with emphasis on vocabulary and memorization and little time for play and exploration. If these children had had a really good chance to play with science materials in the earliest grades, this approach might be all right. But since at least several students in any given class will not have had that opportunity, the upper elementary or middle school teacher who wants to reach the known gifted and help others find their gifts must allow this kind of preliminary play. Teachers may balk and say, "We don't have time for that!" But the important goal is not to present material but to help children learn. If play can be a bridge to learning, we cannot exclude it.

Students learn efficiently and rapidly when they are interested in a subject and motivated to learn it. Yet they can be incredibly slow when faced with something they find boring or uninteresting. Well-meaning teachers often waste enormous amounts of energy trying to teach material students do not find interesting. Far better to let them explore until they are ready to learn that "other content" so dear to the teacher. Students can learn this material in perhaps a quarter of the time when the content is placed before them at the "right" moment. That moment may be different for each student, reinforcing the need for individualization. At the elementary and middle school level, the teacher's job is not to produce research scientists, but to encourage exploration, curiosity, and a genuine fondness for "doing" science.

High school science instruction takes two principal forms. The first is ordinary classroom instruction. The second is independent project work. In some courses these two may be mixed. Attempts to "liven up" the regular curriculum have employed so-called "guided inquiry." Contractual instruction, audiovisual-tutorial methods, and mastery-learning approaches have been the most widespread means of individualization. These approaches do not generally accommodate the need for "play" and truly independent work.

University-level work presents the same problems found in high school science courses. Most instructors assume that playtime is over and solid learning can begin. But they are wrong to assume that most students are already motivated and will work at or near their potential, given ordinary assignments, tests, and papers. There are a few programs that recognize that exploration in an open and unfocused mode will motivate gifted students and enable them to emerge and progress. One used for several years at St. Lawrence University has been labeled "a kindergarten" by colleagues who apparently feel that there are enough talented people around without having to "mine the tailings dump" or look around in the waste pile for any gifted students who lie buried there. They maintain that examinations sort out the already motivated, regardless of their giftedness or lack of it, and that the 75 percent with grades below a solid B are rightfully left to their own devices, relegated to the "average" pile. But many of the truly gifted fall into this category. (48, 49) We cannot afford to neglect these people.

At all educational levels, good science teaching should foster intellectual curiosity.

Dr. Mary Budd Rowe in her book, *Teaching Science as Continuous Inquiry* (2d ed., McGraw-Hill, 1978), helps us to understand good science teaching:

> Science is a kind of journey into the unknown; with all the uncertainties that new ventures entail. Doing science means using intuition; it means creating abstract ideas out of concrete instances in order to find out:
>
> 1. How things work (description)

2. Why they probably work that way (explanation)
3. What must be done to make them happen in other circumstances (control).

Students and teachers must ask these questions:
1. What do we believe at this point in time?
2. Why do we believe it?
   a. What is the evidence?
   b. Where did it come from?
   c. How did we get it?
3. Is there another way to interpret it?

### Activity: "Whaduhyuh Know?"

Ask students to describe their background in science, what topics they have "covered," and how many times they have studied various material. Whenever students mention a new topic, stop for a moment and ask students to discuss whether this is the first time they have thought about this idea. If not, where have they heard about it—school? TV? radio? books? newspapers? magazines? How much do they know about it? How interesting is it? Do they want to spend more time on it or should it be left aside?

## SCIENTIFIC LITERACY AND THE GIFTED

Much of what is written about teaching the gifted in science focuses on the gifted student who might one day end up in a science-oriented career. Too often, science has been considered the special preserve of professionals rather than a normal human interest that all people can share. The increasing complexity of our technologically oriented world makes it imperative that every person be literate in scientific matters. Voters will be called upon to make choices based on an understanding of the consequences of scientific and technological work. Problems posed by energy, nuclear power, medicine, drugs, conservation, forestry, agriculture, nutrition, health—all require clear and inventive thinking that takes advantage of useful scientific approaches and viewpoints. Thus, programs for gifted students will serve those students better if they contain elements that encourage the consideration of environmental and social issues. Moreover, an awareness of one's world can provide a lifelong source of extraordinary enjoyment and recreation for every person. Any balanced program for gifted students needs to incorporate opportunities to learn about, appreciate, and value the natural world and develop a powerful and sophisticated awareness of how things work and of natural processes, limitations, and possibilities. The ultimate goal might be to see every learner want to spend time in active pursuit of "scientific understandings."

## SCIENCE AS "SERIOUS BUSINESS" OR "PLAY"?

At its best, science is pursued the way small children play. Children try something, and when it doesn't work, they just try it again, or try it another way, or introduce some variation. Instead of thinking of "failure," their attitude is just, "Oh! That didn't work. Now let's see what happens when we try *this*." Without getting frustrated, children can apply dozens and dozens of permutations and combinations to a problem they are working on in their play.

The best scientists keep their work in perspective and do not take it too seriously. Such detachment is essential to exploration and quality. If science is treated too seriously in the classroom it can become very dull and boring. Science requires a quality of playfulness. When one is seeking connections and relationships, many avenues of exploration are bound to be dead ends. If one took each of those "failures" seriously, depression would surely result.

## COMPETITION AND GIFTEDNESS

Americans thrive on competition. According to one prevailing idea, schools must foster competition among students in order to prepare them for a world in which everyone is out to win, but most will lose. The existence of prizes in science (the Nobel prizes and the hundreds of others that are equally or less prestigious) supports this idea of winning. Such prizes fail to take into account that great discoveries are never made by a single person acting alone. Science is a collective, cooperative enterprise. The last person to place the final piece in a jigsaw puzzle could never accomplish this had not the rest of the puzzle already been assembled. The history of science is replete with stories of several people working independently, yet coming up with a major discovery at about the same time. The concept of evolution that serves as a cornerstone of modern biology was "discovered" not only by Darwin, but at exactly the same time and in essentially similar form by Wallace. Darwin was just a better publicist. Consider an Olympic contest in which everyone remembers the name of the gold medal winner, but all soon forget the name of the person who came in a fraction of a second later. Is the runner-up really less skillful than the person who happened to fall across the finish line first?

One unfortunate result of the competitive drive has been increased falsification of scientific data. A good deal of cheating has been uncovered recently, even among supposedly distinguished scientists rushing pell-mell for the finish line. (44) Industrial firms have introduced more big prizes for which scientists can compete. Not surprisingly, there is a strong movement among many scientists against further commercialization. This is bound to increase competitive feelings, increase cheating, and diminish the disinterestedness which is an essential quality of good science.

At the same time we see an increase in competitiveness, we are also seeing more science done on a cooperative, teamwork basis. Today's problems require the use of sophisticated technology and the combined effort of many specialists working together. This rise in teamwork and the diminishing of the old "star" concept has been one of the most significant recent developments in science.

Competition often surfaces in the schools, sometimes with dire effects. Science fairs are one manifestation of this competitiveness. The gifted are expected to come up with elaborate science projects that bring the same kind of glory to their schools as does athletic achievement. Science fairs should be a place for sharing knowledge, serving the same important function as scientific professional meetings. It surely seems that science fair prizes are inimical to the growth of student potential and destructive to overall programs for the gifted and to "science" in general. By the time students reach competitive college science programs, especially premedical programs, it is not uncommon to sabotage each other's work in order to get to the top of the pile. This is an extraordinarily destructive situation.

Teachers of gifted students should emphasize science as a cooperative function. Parents and administrators may pressure teachers to push for the science fair prizes, but in so doing, teachers stand a strong chance of inhibiting and destroying giftedness rather than nourishing and developing it.

### *Activity: Competitiveness Versus Cooperation*

If you give tests in class, try the following procedures for a test: Tell the students that you only want one paper turned in for the whole class and that every student will be scored on the basis of that paper. Alternatively, separate the class into two groups. Have half of the people take the test individually; have the other half work as a team toward a group score. Have the two groups compare the "best" of the individual papers with the group paper.

Ask students to teach each other science information. Divide the class into groups of four. Assign different material to each student in the group. After each student has mastered the material assigned, have the student teach other group members. Quiz students individually, but tell them their grade is determined by the group's average. Afterward, discuss the process and its implications.

Ask students to read about cheating in science and discuss this problem. Some suitable papers are those by St. James-Roberts (44), Gillie (13), and Wade (53).

# USING PROGRAMS FOR THE GIFTED AS ROUTES FOR CHANGE

In general it is difficult to bring about changes in U.S. schools today. In most cases the "regular programs" are so well entrenched that no real changes have occurred for many years. During the 1960s the National Science Foundation invested several hundred million dollars in curriculum reform projects to touch all the sciences at all levels from elementary school through early college. Now, only a few years after the last of these projects completed its primary work, a visit to most science classes shows that science is being taught in essentially the same way it has been taught for the past 50 years. Schools are extraordinarily resistant to change. There have been changes in emphasis and in subject content, but few changes in basic patterns of authoritarian classroom instruction.

Even though there is some danger in labeling individual students as "gifted" and thereby singling them out as more special than their peers, programs for gifted students can provide routes for change. As such, they warrant support.

All people are gifted in some area of *science*—if one accepts a broad definition of science. If one accepts only a narrow definition, one would also have to exclude large numbers of science "heroes," as even a casual look at the history of science would show. (6, 54, 18)

The specific activities suggested in this book can be applied successfully with any students, whether or not they are labeled as "gifted." These activities involve a large amount of individual, project-oriented work, and such work has always been acknowledged as appropriate for "bright" students. Some of the more unusual games and approaches presented here are especially appropriate for creative people. Teachers who believe that giftedness is latent, somewhere, in everyone can introduce these activities in any or all of their classes.

# GIFTEDNESS AND ADVANCE PLACEMENT IN SCIENCE

Many school administrators and parents mistakenly believe that conventional "advance placement" courses provide a satisfactory vehicle for all gifted students. In a majority of cases, these programs designated for students gifted in science end up as attempts to bring first-year college science content into the high school, complete with college-type lectures and college-type labs. These advance placement courses cater to those students who are "course-wise" and know how to give content-bound teachers exactly what they want. The courses do little, however, to help develop the special qualities of creativity and inventiveness that are the mark of giftedness in scientific work. The programs largely stifle students rather than provide an environment for growth. Most university science instructors would prefer to work with students who have not become intellectually bound up in advance placement courses. With such students, the teacher's first job is to free them up to think creatively. If teachers who are seriously interested in helping gifted students are assigned an "advance placement" group and given a college textbook with which to work, they can use some of the procedures suggested in this book to convert a standard advance placement program into one that can keep the creative spark alive and still deal with "appropriate content." The primary focus needs to be on the asking of questions and not on the finding and "learning" of answers. One approach is to convert standard placement sections into genuine research-based seminars.

# THE LEARNING ENVIRONMENT

In any program for gifted students the general feeling of the educational environment as a whole is more important than the particular curriculum and materials. *How* things are done is of greater importance than *what* things are done. Above all else, the learning environment must be *nourishing*.

An essential quality for the classroom is what Bertrand Russell (42) characterizes as friendliness: "What is wanted is neither submissiveness nor rebellion, but good nature, and general friendliness both to people and to new ideas....If the young are to grow into friendly adults, it is necessary, in most cases, that they should feel their environment friendly."

In schools that already have special programs, teachers may work with groups of students who are specially designated as "gifted." It is important to help these gifted students find out that other classmates are also "gifted." They need to recognize that all of their peers are special, but in different ways. Students need to feel good about their own gifts and comfortable in letting other people see them exercise these gifts. At the same time, they need to develop an appropriate degree of humility so that they recognize that everyone participates and is able to make contributions, each in the way he or she can offer most. The group should be led to realize that every human being is unique.

## THE GROUP AND THE INDIVIDUAL

The American insistence on the "efficiency" of instructing students in groups is based on the questionable need to "cover ground." This may be one of the single most harmful attitudes that has long hindered the growth of gifted students. Students learn best when they are motivated to learn, and individual students are motivated to learn different subjects at different times. A first essential aspect of a nourishing learning environment is that when students become interested in a subject they must be allowed to continue to follow these lines of interest on an individual basis or in small groups. Later there must be ways of allowing them to rejoin the whole group. At any point in time, the teacher will have an ongoing sequence of activities in progress. But individual students or groups of students may "get off the train" whenever they wish to make a "side trip" or, in Renzulli's terms, "conduct an investigation." (51) At times, the majority of students in a group may be on their own side trips. When this happens the teacher's role becomes that of facilitator, helping students to complete their own work. Usually the teacher's role changes back and forth between that of generator of ideas and leader of activities and that of helper, assistant, facilitator. In both roles the teacher has a responsibility to provide feedback, ideas, and evaluations when students request these.

There is another reason for allowing students to opt out of activities suggested in this book. Teachers who try to force students into awareness exercises, meditations, creativity games, and other special activities lack sensitivity to student needs and feelings. These exercises have psychological ramifications. Misuse of standard, cognitively oriented exercises can be destructive, but misuse of exercises that make use of perceptual skills and abilities can be an invasion of a student's privacy. Students must participate voluntarily in such exercises or not at all.

Teachers should recognize, too, that a student who wants to work independently at a given time does not necessarily wish to avoid a particular exercise. It may simply reflect the direction of the student's energy and interest at that moment. Teachers wishing to work effectively with the gifted must be sensitive to their students' needs.

## AN OPEN-ACCESS EDUCATIONAL ENVIRONMENT IN SCIENCE

An environment that fosters giftedness in science should provide open access to resources and ideas. This does not necessarily mean a big open room without walls, although that might be helpful. It does mean maximum freedom for students to use resources at any time. Science often involves the measurement and study of "things." The environment must provide a wide range of things left out in the open most of the time. Cupboards and shelves should be open so that all resources are easily and quickly available.

An open environment also implies access to the school library, to other teachers (with the understanding that this requires their permission first), and especially to the outdoors. Regular access to the schoolyard or arrangements for walks into town or for field trips may be difficult in some schools, but a teacher of gifted students should be ready to apply continuing pressure to gain this access.

The materials provided depend to a large degree on the particular "discipline" the class is studying. Almost anything goes for teachers in the elementary grades and in so-called "general science" classes through most or all of the middle school or junior high school years; teachers need not feel restricted by traditional content boundaries. As it works out, teachers in elementary schools generally have limited funds and must make do with more ordinary, everyday items. Secondary teachers can generally afford more "scientific" equipment.

A wide variety of excellent curriculum materials especially suited to gifted students is available as a result of the large curriculum development projects supported by the National Science Foundation over the last 20 years. Every area of the science curriculum has been affected by these projects. Unfortunately, as a result of their emphasis on materials rather than on teacher-student relationships and student learning needs, these projects did not have the great impact they might have had. Nonetheless, they provide a remarkably useful resource bank of materials and ideas that can easily be transformed for use in student-centered, problem-centered programs. Detailed information about the availability of these materials (including extensive teacher-preparation materials) can be obtained from the National Science Foundation, Washington, DC 20550.

Whenever possible it is advisable to use funds to purchase single items of many different kinds of equipment rather than be lured by some "package deal." When possible, avoid buying 20 or 30 expensive microscopes or balances. Be satisfied with a couple of microscopes or even with hand lenses or simple magnifying glasses. Use the remaining money to buy a wide variety of materials. Often the "official" science kits sold by commercial suppliers contain materials of poor quality, sold at high prices. Examine these kits to see what they contain and then visit the local hardware store or junkyard for better and less expensive substitutes. If possible, try to arrange for a "contingency" budget that allows purchase of items from local merchants during the school year. This is better than stocking up on advance-ordered material from science supply houses. Such "stock" materials often gather dust on classroom shelves. Each year's new materials should enrich the total environment so that possibilities grow as time goes by.

Another way to help stock the classroom-laboratory is to ask gifted pupils to act as "purchasing agents." Once students decide what they need for their activities, let them help acquire needed materials. This lessens students' dependency on teachers. Gifted students understand that the questions they can study and the precision they can obtain depend upon the equipment they use. If money for equipment is not available, students can improvise and innovate. Inventing balances and magnifiers demands more sophisticated skills and greater involvement than going to the shelf for a standard item. Pickle jars, old pots and pans, and junk provide countless possibilities for much scientific experimentation. Letting students participate in organizing and equipping the laboratory makes them full partners and opens doors for the development of unexpected talents.

### Activity: The Equipment Game

> An equipment game can help both teachers and pupils get past the objection, "I'd love to study that, but I just can't afford (or get) the equipment we need to do it." Have students make individual lists of all the ways they can think of to use a hinge, a paper clip, an old automobile, or other common object. First students can do this individually without talking to each other. Then ask students to combine all of the lists, keeping track of how many repeats there are. "Opening a door" is not a very imaginative way to use a hinge, but "wearing it for a hat" would be unusual. After the initial listing, the assignment can be expanded by saying, "Now let's list all of the craziest ways to use a hinge."

Often people begin by giving conservative answers, out of fear of being laughed at. However, when they learn that the most unusual answers count the most, their fluency increases. Ability to perform in this kind of exercise is not a "given" gift. Fluency and skills can be learned. A group that cooperates can always come up with many more ideas than any individual or even groups of individuals working separately. In the group, one "silliness" triggers another. And every now and then a really interesting idea emerges.

So far the game may not seem relevant to a science class. Remember, though, that the goal was to figure out how to get equipment for the lab.

> Allow sufficient time for play, even a couple of periods if students get excited by the game. An important rule is never cut off a good game too soon, and never let it drag on if interest slackens. A new focusing element can be introduced into the game: "All right, now I want you to think of all the ways we can use that hinge in an experiment for which we haven't been able to get the equipment." In think-tank terminology this approach is known as "force-fitting." (15, 29, 35) Once students have become fluent in generating many different kinds of ideas, they can then focus on a direct problem of the lab. If the hinge doesn't produce any useful ideas, move on to another item: "All right, what other items might we think of that we could use instead of a hinge?" The teacher, although acting as leader, should participate whenever possible. Perhaps a student might be asked to be a "leader" or "recorder" while the teacher joins the group as an idea generator.

Creativity games like the previous one profit from a device on which ideas can be recorded. Most teachers let classroom activities revolve around chalkboards and overhead transparencies. The trouble with chalkboards is that good ideas get erased before long. A pad of newsprint or butcher paper provides plenty of cheap sheets on which ideas can be retained for further examination. Science is mainly a game of connections, a chance to see that things are in some way related to each other. The recognition of unusual and unexpected connections is called discovery, and out of discovery come major advances in both learning in general and in science in particular. Jean Piaget has said, "Nothing is ever learned except to the extent to which it is discovered or rediscovered." (38) If ideas are kept available for inspection, students are more likely to make associations. During a session with gifted students, tack up sheets of newsprint covered with words derived from games and discussions. Some of these words and ideas may call forth an idea directly related to the "science problem of the day." After an investigation has been completed, students can review the sequence of ideas by looking back through the newsprint sheets.

### Activity: The Bag of Tricks

Have each student bring in a "bag of science tricks"—a collection of exciting things they like to do, things related to science. Leave the bags somewhere in the room and open them at random to present as "gifts" to the class. Later combine all the tricks. Add your own tricks. When a student is bored and looking for something to do, allow the student to work on a trick from the bag. Keep the bag going over the school year and through successive years. Each addition to the collection may help some future learner. You can ask that tricks be related to a given study area if your class has a specific disciplinary focus. Better still, ask the class to "force-fit" whatever tricks they come up with to a given task or assignment (see p. 37).

## PROBLEM-CENTEREDNESS AND STUDENT-CENTEREDNESS

All science is interdisciplinary. It wasn't until 200 to 300 years ago that people began to separate biology, chemistry, physics, and other areas. Now the differentiation has gone so far that most scientists understand little of what colleagues in even adjacent fields are doing. As soon as someone starts to do work in one of the "cracks" between two areas, a new area of specialization arises, followed soon after by a new professional journal.

Despite the degree of specialization, there is a great deal of overlap between these specialties. In fact the early training of all scientists covers much the same ground. A high school student taking biology, chemistry, physics, and earth science will find that as much as a quarter to a third of the subject matter is repeated. Students keep going over the same so-called "fundamentals." By the time science-oriented students have completed twelfth grade, they have probably been over the same material, at the same level of sophistication, as many as five or six times. Small wonder that so many students—especially the bright ones—reach college already bored and "turned off" to science.

Small children draw no distinctions among scientific specialties. They ask general questions that pay no heed to boundaries. Gifted students tend to do the same thing. They ask, "How does this work?" rather than limiting their questions to the ones permitted in a fixed discipline. This is the difference between *problem-centered* science and *method-centered* science.

Most science as written in textbooks and taught in schools is method-centered. First the student works through the fundamentals of biology, learning its methods and its subject matter. If an unrelated question comes up (concerning soil, the atmosphere, chemistry, or rainbows), it is usually disqualified as being off the subject: "You'll learn about rainbows next year in physics."

Gifted and talented students are often frustrated by a method-centered approach. Those who are well "socialized" may be content to wait until later to deal with the questions that really interest them. But many children, including many intellectually gifted children (50), simply "tune out" of class, occasionally coming to life when by chance something interesting captures their attention. Most working scientists tend to be method-centered, too, since they've been trained to think in that manner. Faced with a problem outside of a particular field, the method-centered scientist might either farm it out to some other expert or ignore it completely. This despite the fact that in science the most significant answers to questions often come from an adjacent or even distant specialty rather than from the investigator's own garden patch. Most of the really interesting and vital ideas are hybrids resulting from cross-fertilization. That's what "hybrid vigor" is all about.

Problem-centered science encourages students to ask any question that interests them. Teachers and students can use any method to work out the problem. Of course, time is spent looking around in search of ways to approach the problem. At first glance this might seem like an inefficient way to proceed. How can you work on a problem when you don't know how to proceed? This attitude leads people to the same "safe" questions over and over again, questions they know will lead to satisfactory and acceptable answers. This is the low-risk approach to science. It produces few new insights and little excitement or discovery. In contrast, working in a problem-centered manner is fraught with the peril of possible "failure," but the potential rewards are enormous.

Many of the most gifted students work especially well in a problem-centered mode.

The teacher who has worked mainly in method-centered modes may find it frightening to enter the world of problem-centered studies. Teachers can expect to be "out of their depth" often. They become coinvestigators with their students and will less often be "one up" on them. Of course, the teacher has several advantages as a result of methodology, life experience, and maturity. Teachers working in a problem-centered environment must be ready to learn from the provocative questions of gifted six-, eight-, ten-, or eighteen-year-olds. Teachers also must be ready to have students take a great deal of responsibility for finding out things on their own. In a problem-centered mode, students may be asking questions that cannot be answered as yet. Teachers should expect *not* to know the answers to many of their questions and, more important, perhaps not even to be able to find out the answers. In a problem-centered environment, teachers must be ready to say, "I don't know, and I don't know whether anyone knows. Why don't *you* go see if you can find out for us." The responsibility for learning must be passed from the teacher to the students. The teacher no longer serves as an encyclopedia, although the teacher does not, of course, deny information. The teacher becomes an equal participant in the exploration of science.

### *Activity: Finding the Problem*

Have students take any activity from the textbook or laboratory book. If the exercise focuses on biology, ask them to find the physics, chemistry, and earth science hidden within it. (For the other subjects do the same.) Now have students find the literature, music, poetry, and social studies hidden in the science exercise. If they have trouble completing this task, ask them to modify the exercise as necessary to add these dimensions.

Have pupils keep track of class assignments and textbook assignments to see whether these begin from a real problem or question or whether they simply provide practice of a skill or technique. Gather science reports from the newspapers and popular magazines, and have students rate these as to whether they are centered around a problem or are confined by a method. Are the newsworthy reports "bound" by interdisciplinary lines? What differences are there between sciences as practiced and science represented in textbooks?

### *Activity: Experiments Always Work*

Go beyond the ideas suggested in textbooks and laboratory manuals. Ask students to speculate and write down their thoughts about the experiment.

> There is no such thing as an experiment that didn't work! Analyze the outcome and ask, "Why didn't the experiment do what we expected?" and "What would happen if...?"

If I am truly interested primarily in the growth of the learner, rather than setting myself up as a defender of my scientific profession or a gate-keeper of the "standards" and "certificates," I understand that no item of information or no skill is so important that any certain learner needs to learn it on any given day of any given year. I also understand that no one piece of learning need necessarily precede any other or be prerequisite, so varied are the backgrounds and mental processes of different and unique human beings.

No matter how hard teachers try to make their classrooms "student-centered," they are bound to try to "lead" their students from time to time. Because teachers are more experienced and know how some aspects of knowledge are stacked together, they also will want to share their expertise, shortcuts, and approaches with their students. Surely there is room for teacher-centered needs and content-centered ideas. However, teachers should focus primarily on learners and their needs if they wish to foster maximum growth of the gifted.

# SCIENCE ACTIVITIES FOR THE GIFTED

The first sections of this monograph have presented several activities to use in exploring with students the nature of giftedness and the nature of the learning environment. All of these activities involved "scientific content" in areas such as statistics, population studies, perceptual psychology, and human biology. The sections that follow suggest other activities that can be directly applied in work with gifted students. Many of these activities even can be used in classes for the so-called handicapped or "educable mentally retarded," since some of these students often show advanced abilities when given the opportunity to perceive concepts and demonstrate ideas in nonacademic forms.

It is important to remember that the kinds of activities proposed in this book are intended for *voluntary* participation. Many can be counterproductive if students are coerced to participate. All require a higher-than-normal degree of mutual trust between students and teacher and among students. The activities also can be used to build trust. Teachers must judge which exercises they and their class can be comfortable using. If teachers are in doubt, they should talk it over with their class first. Always allow an out for students who choose not to participate. Remember that growth and change usually require departures from what one is accustomed to, and thus involve some risk, uncertainty, and apprehension.

The activities presented here have been divided into three categories: (1) general comments about the teaching-learning environment (preceding sections), (2) activities and exercises on perception and values, and (3) activities and exercises related to the development of science skills and concepts.

## EXERCISES IN PERCEPTION, AWARENESS, AND OBSERVATION

The nature of "reality" becomes an important subject in science programs for the gifted. Science is supposed to deal with what is real. Yet how real is an atom? An atom is a concept, not a physical entity. No one has ever seen an atom before. Atom is only a word, coined to represent something of vast complexity.

We perceive things only within the restricted limits of our own "instruments." We see and hear within narrow limits imposed by the nature of our eyes and ears. We invent instruments that let us sense and infer the existence of other physical things. There are whole areas of strange phenomena for which we have no explanations. Some people are attempting to explore these difficult areas, such as parapsychology, just as other "conventional" scientists try to discredit them. Rockwell (30) describes how respectable scientific journals made systematic attempts to exclude publication of work in parapsychology. (24) This phenomenon is common at the beginning of any scientific revolution. (21, 3) Galileo had to deny his certain knowledge that the earth was not the center of the universe or face probable execution at the hands of the Inquisition. People cling to old beliefs about the nature of reality. Gifted students in particular are likely to challenge dogma. It is essential that teachers encourage the growth of this critical, challenging attitude.

Exercises in awareness can enrich every group of gifted learners. These exercises help students gain new perspectives by seeing things from different viewpoints. Careful observation is one of the traditionally important values of science, whether the subject is physics or archaeology. Questions are generated through observation and because students notice unusual things they wish to understand. Answers to these questions come from further observation and the search for relationships that may explain what is happening.

Athletes never compete without first warming up. The warm-up promotes efficiency and helps the athlete avoid injury. Yet professional scientists rarely warm up their observational skills, and teachers rarely warm up their students. The warm-up is an important skill for gifted people. Many gifted students have a great deal going on in their minds at any one time; careful observation depends on their ability to clear their minds of extraneous thoughts that stand in the way of seeing what they wish to observe.

### Activity: Warming Up

1. Awareness exercises can be done by the whole group or they can be done individually. The body must be relaxed before one can hope for good observation. Thus, before beginning an awareness exercise, it is a good idea to have the group stand and loosen up physically. Students should shake their arms, rotate their heads to loosen their neck muscles, and breathe deeply and quietly for a while. Next suggest that they gaze steadily at some nearby object and study this as carefully as possible. Anything will do: the corner of the room, a crack in the wall, the front desk, the door.

2. Have students assume comfortable positions (even prone on the floor if this is possible in your teaching environment). Once the group members are relaxed, have them close their eyes and lead them through a body and mind

relaxation exercise. The following is typical: "Breathe deeply and quietly....Now tense your body up so that all of your muscles feel tight....Now relax your muscles completely....Be aware of your big toes now....Let them relax just as much as you can....Now be aware of the rest of your foot....Let it relax...." Continue this progressive relaxation exercise until the whole body has been instructed to relax.

3. Hand each person an object or have them select one from nearby: a book, a leaf, a pebble, a pencil. Lead the group through a description of the object: "Notice the stem of the leaf....Really look at the details of that stem—its color, whether it is smooth or rough, whether it is mottled or even in color, whether it is thin or fat, tapering or straight....Now look at the margin of the leaf itself....Be aware of whether it is smooth or jagged....How jagged? ...Look now at the veins that radiate from the center of the leaf...." After each instruction, give students plenty of time to inspect their leaves. At the end of the exercise, collect all of the leaves in a pile, stir them up, and then ask each person to retrieve his or her own leaf. How well have they examined their leaves? Well enough to recognize their own in a pile of similar leaves?

## Activity: *Look—See*

Give each student an object pertaining to what is being studied (plant, animal, rock, star chart, crystal). Ask each student to spend a whole period writing down all that they see in the object and all that it makes them think about. The teacher can also participate! No questions should be asked or answered. The task is intended to force people to observe.

## Activity: *Feel*

Mark a number of walnuts with different colored magic markers (one dot will do). Ask each student to take a walnut and explore it. Ask them to feel it with their fingertips and discover what makes this walnut unique. Place the walnuts in a brown paper bag, and without looking inside, ask each student to find his/her own walnut. (The walnuts can be identified after the activity by the colored dots.)

## Activity: *Sound of Nature*

With a tape recorder, find and record human-made natural sounds. Ask students to identify the sounds. Compare these with natural sounds. Ask students to describe how the sounds make them feel.

## Activity: *WIGO? (What Is Going On?)*

Ask students to stop whatever they are doing and notice in detail what is actually happening at that moment in the classroom or learning environment. Have them list these events on sheets of newsprint. Discuss these events.

Bois (2), who invented WIGO, points out that a person brings to any task a load of "baggage" from outside. If that person is to deal effectively with that task, he or she must first be aware of what is going on (WIGO) both within and outside the person. This awareness makes it possible to set that baggage aside and move ahead. WIGO can help people regain the spirit of the fascinated child who pauses in the middle of a walk to stoop and admire a tiny blue flower growing among the ivy. Teachers must help students retain or regain this quality of naivete and delight.

Whenever you sense confusion, tension, or lack of focus in the group, just call out "WIGO!" Let this word signal students to stop for a moment, take stock of what is going on, and discuss it briefly (in nonjudgmental terms). Students can then return to other activities with renewed awareness.

### *Activity: Clearing the Mind*

The purpose of this activity is to quiet the mind, reduce preoccupation with extraneous events. Some of the most perceptive observers of nature have been Zen monks and others who chose the quiet, contemplative life. (40, 23) Ask the group to sit quietly for a few minutes and count their own breaths (one...two...three...four...one...two...three..., etc.) or sit quietly and repeat "one" over and over on alternate breaths.

## FANTASY EXERCISES

Inventiveness and the ability to fantasize are other qualities that need to be freed up in students who are usually told "Pay attention!" rather than "Daydream!" By focusing attention on imaginative play, teachers emphasize that this is as much an element of intellectual activity as it is problem solving. Fantasy exercises can provide springboards to more formal investigations in particular areas of interest. Many fantasy exercises have been described elsewhere. (46, 35, 34, 9, 4) An example follows:

### *Activity: Daydream!*

Relax the group with a short relaxation exercise (see pp. 42–43). Now slowly narrate directions for a fantasy excursion: "Imagine yourself as a molecule of water in the middle of the ocean....Get in touch with what it feels like to be that molecule....Be aware of all the other molecules that are around you....You are moving back and forth in a wave....Gradually you approach the shore as you travel along in a current....Experience the feeling of movement....As you approach the shore you feel yourself caught in a wave moving rapidly forward....You are in the white, frothy foam riding on top of the surf....Your wave crashes up onto the shore....You feel the sand underneath you....Now you percolate down into the sand...."

This kind of fantasy exercise could lead into questions and answers, readings, experiments, and other activities related to water and waves. Any science process or object lends itself to treatment through fantasy. Let your students suggest and lead their own fantasy exercises.

## VALUES AND PERCEPTION

Students come to the learning environment heavily loaded with values. Weeds are bad; litter is bad; pollution is bad; conservation is good; natural goods are good. A primary job for teachers of the gifted is to call attention to the fact that these values are culturally determined. How do you encourage students to look at the wonder of a dandelion or explore the patterns and colors of trash in a dump? A rumpled piece of paper in a gutter, when seen for the variety and interest of its folds and texture against a background of rough concrete, can also be an object of beauty. One teacher had students collect litter at Christmas and make a "trashmas tree." Much modern art challenges the dogma of good-bad value judgments by focusing on objects as seen for themselves. Students need to learn to see objects just as they are—just there—and not through value filters. The smell of a skunk is thought of as disgusting because we have been taught to dislike it. If you learn to perceive it as just another smell, the sense of disgust goes away (although the smell certainly does not!). The same is true for food preferences, color preferences, and many other areas of perception. We approach objects full of biases that lead us to judge and categorize. The ultimate problem in science is to try to see things as they are, in what is often called their "suchness"—their value-free essence. We cannot totally escape evaluation and judgment, but by being aware of our attitudes, we are able to decide what we wish to believe rather than merely accept a set of beliefs we have acquired from others. This freedom of observation is essential to scientists, and teachers need to lead gifted science students in this direction.

### Activity: Evaluate!

To emphasize the influence of values on perception, ask students to list the characteristics of several items they strongly dislike and strongly like. Then have them "rate" aspects of their environment in terms of likes and dislikes. Ask them to look at items they especially dislike and try to describe them as things of great value and beauty. Then take objects or ideas that are highly valued and list all their possible bad points. Finally, compare the various lists, looking for differences that are a result of viewing the items negatively or positively.

# MAKING THE STRANGE FAMILIAR AND THE FAMILIAR STRANGE

Once an object becomes familiar, people stop looking at it carefully and just assume that they know what it looks like. Yet everything one sees is, in some way, different each time one sees it. Most schooling involves familiarizing students with things they don't know—"making the strange familiar." When we try to make the familiar strange, we enhance seeing. A favorite technique of forgers is to forge signatures *upside down*. That way they can see each letter as it really looks, and they are not influenced by their own familiar concept of how that letter looks. This is similar to the way that scientists see differences in organisms that lead to the differentiation of new species, types, and theories.

There are other ways to make the familiar strange: magnify an object many times, shrink it drastically, or look at only a part of it. Under magnification a small portion of a butterfly's wing or a section of wood grain looks not at all like the object itself.

Another way to see things differently is to make yourself switch the relationship between what is called the "figure" and the "ground." An object is seen in relationship to the background upon which it is superimposed. No object exists without a background, and no background exists without an object in front of it. This concept is in all scientific observation, in art, and, in a metaphoric sense, in literature.

In observing objects, you can look either directly at the object or at what surrounds the object. The space filled by the object is called "positive space." The space that surrounds the object, completely enclosing it and filling up its open space, is called "negative space." Any time we look at an object we see either the object itself or the space around it. When we switch our perception back and forth, the relationship between figure and ground reverses. Examples of exercises and puzzles are in Adams. (1)

### Activity: Making the Familiar Strange

1. Choose any object under study and ask students to look at it as something strange, something they have never seen before. Have them describe the "new" object.

2. Cut out magazine photographs of common objects. Present these to students and ask them to figure out what they represent.

3. Have each student bring in several fragments of pictures the other students can study. Ask students to guess what the fragments represent.

4. Have students draw a desk or chair placed in front of them as a "model." Now have students draw the space around the model—not the chair, but the *space* around it!

5. Ask students to look at a map of the United States. What do state boundaries remind them of? (Cape Cod is often seen as an arm, Lower Michigan looks like a mitten, etc.) What figures do they see in clouds?

6. Have students copy a drawing, first right side up, then upside down. Discuss the results.

7. Magnify or reduce objects or parts of objects and have students work with these "changed" materials.

## Activity: *Conflicting Observations*

1. Collect optical illusions and feature a different optical illusion each week.

2. Magic Can—Before the experiment, tape a weight to the inside seam of a coffee can. Put the lid on the can. Put the can down on the seam; it will roll. Put the can down on the opposite side and it will not roll. Ask students to explain, without knowing about the weight inside the can.

3. Magic Grape—Take two grapes similar in size and weight. Peel one grape and leave the other alone. Put both grapes into a clear carbonated beverage. The unpeeled grape (heavier) will float. The peeled grape (lighter) will sink. Ask students to observe and tell why this happens.

4. Magic Paper Clips—Fill an empty baby food jar with water to the top. Ask students to predict the number of paper clips that can be added to the water without having the water overflow. Add paper clips one at a time (*Answer:* approximately 30–40).

# HARD EYES—SOFT EYES

The human eye is a fascinating device. It has a very narrow band of sharp focus surrounded by a wide zone of somewhat hazy peripheral focus. We learn to value most what is dead-center in our focus. All eye tests deal with the activity of that central focus. Athletes, bird-spotters, and foresters learn the importance of peripheral vision, which locates objects so that they can then be brought into focus. George Leonard (22) has pointed out the importance of this peripheral focus in "seeing" the world and suggests an exercise called "soft eyes." Because our normal focal vision is "hard," Leonard suggests that after a process of loosening up the body (see pp. 42–43), we massage our eyeballs gently until they feel "soft" rather than "hard" as they feel when we are tense. (Try it if you don't believe it!) Then, consciously avoiding direct focus on any single object in the environment, we open our eyes only partially and move our whole head and body around to "take in the scene." Colors are enhanced and we become much more aware of what is going on around us. This kind of exercise is just one of many that help learners realize what powerful instruments their bodies are.

### Activity: Soft Eyes

After a relaxation exercise, guide students to a state of "soft eyes." Have them circulate around the room with eyes half-closed, trying to keep their eyes soft. Show them various objects while their eyes are soft. Ask students to describe any differences in perception or unusual things they have noticed with their soft eyes.

### Activity: The Visitor from Outer Space

Ask students to imagine that their spaceship has just brought them to a planet they have never before visited. They have only the time of one class period before blast off, and people from their planet will never have another opportunity to visit this place. They are to assume that they have never before seen any of the objects they see now in the classroom or learning environment. Ask them to gather all of the information they can about this new planet, so that they can report back to fellow aliens what this place is like.

We can learn a great deal about our perceptions when we look at the world from a completely different point of view. When we put ourselves outside of our usual perceptual framework, we notice things we have heretofore overlooked. This approach not only helps perception, but also helps us find interesting ways to communicate these perceptions to others. The ability to describe something in detail requires that we see the object *as if for the first time*.

## SCIENCE AND INTERCONNECTEDNESS

Many gifted people are especially good at seeing connections and relationships. (50) This quality, if not already visible, can be developed in people. The television series and book *Connections* (5) is based on the unlikely chains of connections responsible for the advance of science and technology. An important process within science is what has been called "synergy," the coming together of many circumstances, the combined action of which produces innovation, progress, and understanding. Teachers who wish to promote students' ability to see and appreciate interconnections must learn to accept, encourage, and value ideas and processes that initially may seem unrelated to the subject being studied. The exploration and examination of seeming irrelevancies may lead to unexpected insights.

### *Activity: Connections*

Have students prepare two lists of words: one a list of vocabulary and concepts from a science unit, the second a list of words selected at random from a dictionary, from another course, or from a newspaper or magazine. Have them draw a random series of lines to connect words from the two lists and then ask them to invent or find a connection between the two words. Later ask them to find or invent connections relating three, four, five, or more words from the two lists. Can they suggest ways of testing the connections they have proposed?

## EXERCISES IN APPLICATIONS OF SKILLS

Science, as defined in this book, is primarily a process—an activity that people do rather than a subject that they learn. The whole environment that surrounds us is the "content" of science. Thus, everything that a student does or sees can become a subject for study and analysis, a valid point of departure under the rubric "science." Fundamental skills to be developed among gifted students include skills in exploration, creativity, problem solving, observations, synthesis, and the formulation of questions.

### *Activity: Finding the Science Around Us*

Make any of the following assignments: Find the science in your physical education class. Find the science in *Hamlet*. Find the science in your locker. Find the science in your breakfast, your soft drink, your girlfriend, your parents, your walk home, your favorite TV program....

To make the assignment more specific, change the word "science" to "biology," "chemistry," "geology," "astronomy," "physics," "medicine," or any other field of interest.

Repetition of this assignment emphasizes that science is an ongoing activity that touches every moment of every human life, whether or not we acknowledge it.

## Language, Classification, and Semantics

Science instruction continues to be based largely on language. In the Preface we discussed the idea of words as metaphors. Every word in science is only a symbol or metaphor for what it represents.

In one geology class I carried in a big hunk of a rock classified as "anorthosite," a widespread rock-type that has been an object of my research for some 17 years. I told the group, "This is an anorthosite." A student raised his hand and asked me, "What is an anorthosite?" I put the 10-pound chunk onto his desk and said again, "*This* is an anorthosite." Said he, "Yes, but tell me what it *is.*" Again all I would say is that *this* was an anorthosite. He wanted a set of verbal symbols when I was insisting on offering him the whole thing, the real thing, a thing beyond words, the very rock itself. We tend to seek substitutes for things rather than accept the things themselves.

A colleague gave some students a group of thin-slices (thin sections) of rocks and directed the students to describe what they saw. They resisted and asked to be told what to describe, what was acceptable to say. This was one of the hardest assignments they'd ever had in science because they were asked to describe *what they saw* in whatever terms they could, without relying on someone else's directions. In a similar exercise, I challenged a group to spend three full hours describing a single rock sample. They were unable to complete the exercise, however, so bound up were they by their past experiences and by the expectations of previous teachers.

It should be one of the primary goals of teachers to help the gifted begin to see with their own eyes. The development of primary awareness as experienced through their own uses and as communicated in ways that they themselves invent is a prerequisite to the effective expression of giftedness.

Classification has become one of the primary activities of scientists. They have felt the need to group things together to make an apparently chaotic world seem simpler. A principle of general semantics (19) is that every object is unique. That piece of anorthosite I showed to the student was unique when compared to all other pieces of anorthosite—just as every human being is unique. For complete accuracy from a semantic point of view, I would have had to label that piece of anorthosite "$anorthosite_1$." A second piece would be labeled "$anorthosite_2$," etc., with each piece given an identifying number. But life would be difficult if every pebble we picked up had to have a different name and if we couldn't talk about "the raccoons that feed outside our window every night," but had to come up with a different name for each of them.

Preoccupation with classifications instead of with what they represent leads us *not to look* at an object or individual. If we see an object and identify it as a Canada goose, then we are excused from really looking at *that* Canada goose. Once we have assigned a name we are excused from further involvement. Most children have been conditioned to group together objects that look or act in much the same way. Exercises that help show the wide diversity within a "class" can be of great interest to students. The recognition of the *real* diversity within the world is exciting. The discovery of something *new* is an important part of science and often the subject of scientific papers. Teachers have the chance to share with students the secret that each blade of grass differs in visible ways from any other blade of grass that has ever existed. What wonder there is in this kind of revelation!

### Activity: Classifying!

Have each student take 10 pieces or samples of a material or organism related to current study. In biology, use plants, animals, cells, etc. In earth science, use rocks, fossils, maps, folds, models, etc. In physics, chemistry, or general science have them choose a process or a measurement to be repeated. Ask students to find as many similarities as possible in the 10 different samples of the "same thing." Then have them find as many differences as possible.

As students begin to explore similarities and differences among given items in school, they should also be encouraged to adapt this skill to their other environments outside school. Take any 10 objects in the room. Write the name of each at the top of a piece of newsprint. Ask students to write down all the metaphors or analogies that can be used to label this object. Leave the lists up for several days so students can add more and more.

Have students gather several other objects that have the same "name" or label and that look as much alike as possible. Have them find the differences between these objects. Then have them find a large number of objects that have the same name but look as different as possible from each other. Next have them find a large number of objects which have different names but which look as much alike as possible.

## CREATIVITY EXERCISES AND GAMES

Often people equate giftedness with creativity. Yet every person is far more creative than we think, and all people can be inventive under circumstances that honor and foster inventiveness. It has been established by numerous researchers that creative abilities increase with practice, support, and training or instruction. (26a, 15, 28, 35, 52) There are many creativity games (see the sources mentioned above), and the goal of these games is to increase the fluency with which people generate ideas and combinations of ideas when confronted with a "starter" idea. All of the games stress the need for cooperation and support and downplay competition. All require the deferral of evaluation and judgment. Teachers interested in promoting creativity must recognize and avoid "killer statements" (such as "What a crazy idea!") which inhibit the freewheeling and playful generation of ideas. Later there comes a time to select and elaborate on a few ideas taken from a large pool of possible ideas, but a long period of play precedes this stage.

This notion may contradict instructions found in many "teaching methods" books which tell teachers how to sidestep any "irrelevant" questions and ideas brought up during a discussion. These "traditional" methods suggest subtle and not-so-subtle ways of manipulating discussion to follow the teacher's lesson plan. But teachers who wish to promote creativity in the gifted must listen to each idea contributed, no matter how remote it seems at first. The teachers' task is to find the relatedness of the response.

### Activity: *Looking for the Bright Side*

One approach to finding the "germ" of every idea is to restrict your own and student responses to positive comments. Before beginning a discussion say to students, "Tell us the things you like about every idea that is proposed." Give each student a point for every positive thing he or she can find to say about an idea. Subtract two points for every "killer statement" (a statement that ridicules, criticizes, puts down, diminishes, or adds a negative element). The person with the greatest number of points wins. Or have students propose the stupidest, dumbest, course-related idea that they can. Then take this "stupidest" idea and find as many good points as possible using the point system described above. Or have students propose some truly worthwhile course-related ideas. Now find bad things to say about one of these ideas. (Reverse the point system: now "killers" win.)

### Activity: *Synectics*

Synectics, Inc., of Cambridge, Massachusetts, suggests the following approach. When introducing a new topic to students, write two questions on the chalkboard: "What I know?" and "What I want to know?" Have students begin to answer the first question, listing the responses in a column under that question. As responses diminish, have students answer the second question, "What I want to know?" listing their responses in a column below the question. As students answer the question, they may discover additional answers to the first question. These also should be listed on the chalkboard.*

---

*Synectics, Inc., Cambridge, Massachusetts.

A focus on positive responses brings out the best and most imaginative ideas as students lose their fear of being ridiculed, challenged, or evaluated. Competition can hurt, and the use of a point system in this activity can be counterproductive. Try the activity without the competition!

## ILLUSTRATING IDEAS THROUGH OTHER MEDIA

Singing, writing poems, drawing, play-acting, and reading nonscientific material are not usually thought of as science-related. However, all disciplines are interrelated. Because these and other activities expand a student's viewpoint, they can help generate scientific ideas, ease the development of scientific skills, and promote the learning of science content. Encourage students to pursue these activities, as suggested by the following examples.

### *Activity: Sing-a-Song*

Ask students to listen to songs with science content. If possible, have them bring in recordings of these songs to play to the class as a basis for discussion. Biological and earth-related subjects may be easiest to find, but ask them to "stretch" to find chemistry and physics. (Some classical pieces such as Holst's *The Planets*, and Grofe's *Grand Canyon Suite* are obvious choices. Learning to "stretch" and "force-fit" can be more interesting, however. "Geographic" songs are common (e.g., "Old Cape Cod," 'Oklahoma," "California Earth Quake"). Students might also enjoy some of the humorous science-related music of Tom Lehrer. Ask students to compose their own science lyrics, setting the next day's lesson to music. (Setting facts to music or poetry can be a useful device to help students memorize science information.)

### *Activity: Bake a Cake*

To review for a test on cells, have students make edible cell models—cake with icing organelles, pizza with mushroom mitochondria, or amoeba cookies.

Franck (12) suggests the use of drawing as a way of "seeing." Drawing and sketching can help science students get a better sense of how items are related to each other. The object is not to create the loveliest drawing, but rather to understand.

### *Activity: Draw-a-Picture*

Have students draw cartoons and rough sketches of what they observe. Have an "art show" (no judging or prizes) of the concepts in a day's lessons.

Include sketches of things observed as well as metaphoric drawings that illustrate concepts.

Have students look for the scientific content in paintings seen in an art museum or in an art history book. Seek both "intentional" science content (of which the painter must have been aware) and "unintentional" science content. Do the same for magazine photos.

To demonstrate the students' understanding of laboratory safety, have them design posters without words that illustrate a safety rule to any non-English-speaking person.

### *Activity: Play-Acting and Reading*

Invite students to act out natural processes: "Some of you be the ocean, some be the beach, some be the waves. Now act out the surf rolling onto the shore." Students might enjoy acting out a nuclear chain reaction or a migration of Canada geese. Exercises that involve body, mind, and the emotions can have maximum impact on growth and learning.

Ask students to find the biological aspects of any piece of literature: for example, Hamlet or the writings of Dr. Seuss or Mark Twain. Or ask them to discuss the scientific realism of the James Bond stories.

### *Activity: Metaphor Game*

List the most important vocabulary and ideas covered in a lesson, a unit, or the whole course. Ask students to develop an extended metaphor for each word or for a certain number of words in a longer list. Types of metaphors include poems, plays, pieces of music, essays, and puns and witticisms. Use the metaphors submitted as a basis for discussion.

Metaphors can reveal a new dimension or provide a new viewpoint and can sometimes illuminate a concept, word, or idea better than a simple definition or direct statement can. Many major scientific ideas have originated in the minds of their creators as metaphors, as "streamlined" models which were later translated into technical, "scientific" language. One outstanding example is the double helix (DNA) of Watson and Crick. (54)

## TV—IF YOU CAN'T FIGHT'EM, JOIN'EM

Much has been written about how television destroys children's minds. Postman (27) believes that television now constitutes the major educational enterprise in the United States and calls it "the first curriculum," relegating schools to second place. Teachers just cannot compete with television when it comes to presenting information in dramatic, exciting ways. Television, with its attractive graphics and its ability to go "on location," has incredible potential for conveying information about distant places, complex processes, and rare specimens.

### Activity: Finding Science in the "Soaps"

Suggest that students look for science-related information on television. Once their "science antennae" are up, it's amazing how much information can be retrieved from even the "worst" of soap operas, cowboy films, cartoons, or situation comedies. An excellent example to use is the Roadrunner cartoons. Featured in many of these cartoons are the laws of motion and unique examples of simple machines. Recommend that students also watch quality science programs such as "Nova," specials produced by the National Geographic Society, and others.

## MENTAL MAPS

Almost all knowledge exists in some kind of spatial framework. Maps help us visualize relationships between people and things. In addition to highway maps or topographic (contour) maps, there are geneology charts, organizational charts, and gene maps. Virtually all information lends itself to some kind of map representation. The following exercise helps students formulate various conceptual maps. Have students draw these maps and then use them as vehicles for discussion.

### Activity: Maps in Minds

Draw a conceptual map of "where you are coming from" today.

Draw a conceptual map of your home town. (Do not restrict the dimensions of the town they choose to represent.)

Draw a conceptual map of how to carry out a laboratory activity.

Draw a "smell" map of your classroom, from memory.

Draw a map of your school, indicating places where you feel good and bad.

Draw a conceptual map of the United States.

Draw original conceptual maps of molecules, atoms, and other things you have been told exist. Do you believe these maps?

> Ask students to write a set of detailed instructions that tell how to get to some point in the school building, school grounds, or town. Then ask them to draw a map based on the written instructions. Turn the students loose and see if they can find their way using the instruction map.

Further details about mental mapping are available in Gould and White (16) and in Downs and Stea (7). Students might like to discuss a point about which the general semanticists (19) continue to warn us, that "the map is not the territory." Maps are representations, abstractions, and symbols, just as words are. They are only metaphors and must not be mistaken for the places or things they represent.

## ANSWERS AND TRUTH

Pablo Picasso once said, "An answer is a lie that makes us feel we've heard the truth." In most science classes the general focus is on getting answers. Yet many science historians agree that there is no sure correlation between "truth" and the scientific answers in vogue at any point in time. Kuhn (21), in a statement that bothers many scientists, points out that when scientists' beliefs shift from one major theory to the next in the supposed evolution of knowledge about an area of interest, it does not necessarily constitute a shift toward "truth." In the complex universe we inhabit, "truth" is relative, whether we like it or not. Our perception and understanding are limited. Imagining that we can really find "answers" and understand "truth" is not only futile, it also is impractical. Better to admit these limitations and help students recognize and accept them. Rather than pretend that we can become omniscient, we are free to hypothesize, to accept science as "another human game," to enjoy our work without taking it too seriously, and to make greater progress toward some ultimate truth, if such a truth exists.

When teaching science to the gifted, by shifting our emphasis away from answers, we are better able to focus on the process of *questioning*. If there are no final answers, students can question without fear of getting "wrong" answers. Open-ended questions leave us free to *wonder*.

### Activity: Questions, Questions, Questions

> Give the class any stimulus (e.g., object, process), and ask students to spend a whole period asking questions about that stimulus, with the understanding that no time will be spent on seeking answers. Such an exercise allows students to express the extraordinary, far-out curiosity that's a hallmark of gifted learners.

### Activity: *Would You Believe...?*

Take a passage from any conventional textbook and have students read it aloud, placing a question mark at the end of every sentence. Use these exercises to begin a discussion of the nature of questions and answers in science. What is it really possible for us to "know"?

The following chart suggests another topic for discussion. It is estimated that throughout history knowledge has doubled in the following pattern:

| | | |
|---|---|---|
| 6000 B.C. –1650 A.D. | —7,650 | years |
| 1650 – 1800 — | 150 | years |
| 1800 – 1900 — | 100 | years |
| 1900 – 1950 — | 50 | years |
| 1950 – 1960 — | 10 | years |
| 1960 – 1967 — | 7 | years |
| 1967 – 1972 — | 5 | years |

(*Source:* Robert Thompson, Union Graduate School)

What do these figures really mean? How could they have been derived? What are the implications of this chart for science, questions, answers, and truth?

## AMBIGUITY AND UNCERTAINTY

Teachers have been advised to make assignments as clear and unambiguous as possible. In that way they can be sure that students will work toward a limited range of "right" answers. But with gifted students, we are looking for a wide variety of highly diverse responses. If unambiguous instructions tend to generate highly *convergent* responses, then perhaps ambiguous instructions and questions will elicit *divergent and creative thinking*.

### Activity: *Divergent Thinking*

Typical "ambiguous instructions" used in the Environmental Studies Project (8) are represented in the following examples:

- Figure out what kinds of things people expect you to do without asking permission.
- Go outside and map something you cannot see.
- Go outside and find a million of something and prove it.

- Go outside and do something you really want to do.
- Go outside and find a change that is predictable.
- Find who in your community are predators and who are prey.
- Go outside and find two things, one of which is responsible for the other.
- Go outside and locate the three most biological (or geological or astronomical or physical) things you can find in the environment.

To extend the last suggestion, we once asked a group of students to take Polaroid photographs of the three most "ungeological things" they could find. Later we gave these same photos to another group, telling them that the first group had taken these as representative of the three most "geological things" they could find. The second group had no trouble at all in finding geological factors in the supposedly "ungeological" photos. Try it.

Ambiguity is one of the certainties of any scientific work. Yet in schools we typically search for clarity and give the impression that the world is well organized and unambiguous. If we acknowledge this ambiguity and play with it through exercises such as those suggested above, we can help students understand the scientific process and the nature of scientific results.

A department chairman in a geology department where I once taught told me I should focus more on what we know rather than "confusing" the students with stories of all of the uncertainties of science. But it is within these uncertainties and ambiguities that the excitement lies, and it is this excitement that can stimulate the creativity of the gifted student.

## STEPPING OUTSIDE THE SCHOOL ENVIRONMENT

Any classroom or laboratory is an impoverished environment for learning science, especially for gifted students. Whatever you can do to arrange observational and other research work outside the school will probably be a strong motivator for gifted students. The ES materials (8) mentioned in this book usually begin with the phrase: "Go outside and...." Excursions need not necessarily be elaborate field trips. Field trips are helpful, but many teachers are paralyzed by the planning and logistics of such trips. They fail to recognize the many possibilities immediately adjacent to the school, whether in the inner city or in a suburban park. Museums, galleries, factories, and businesses are other excellent places to visit. Additional suggestions on how to organize trips for gifted students can be found in Romey. (38, 34)

Local internships and work-study programs can also provide great opportunities for gifted students. If these alternatives cannot be arranged as a regular part of coursework, teachers can often help students find such opportunities on an extracurricular basis. Participating students can bring back to the classroom accounts of their experiences. Such sharing helps enrich the science program.

# WHAT ABOUT THE BASICS?

Everywhere in education we hear about "the basics." Yet we have mentioned this word only in passing. What about teaching essential content and essential skills to gifted students?

In a well-functioning learning environment, vital skills will be learned as an integral part of normal activities. If the skills are "basic," learning them is more or less inevitable. Anything "fundamental" cannot be overlooked because it forms part of the fabric of everything else.

When you isolate basic skills from their context to focus on the skills themselves, learning them becomes deadly dull. Students turn off their attention. Focusing on the basic skills is probably the surest way to alienate students, especially gifted students who can easily differentiate between sense and nonsense.

By taking a problem-centered approach to science content and by encouraging creative behavior, playful attention, and student self-responsibility, we provide the best possible environment for students to master basic skills. We need to encourage but not to force them to write, calculate, reason, analyze, and synthesize. Our primary focus must be on the interests students bring with them to the learning environment. By giving careful attention to student interests, learning styles, and special gifts, we provide them with the best chance to emerge with the basics well in hand.

# BIBLIOGRAPHY

1. Adams, J. L. *Conceptual Blockbusting.* New York: Stein and Day (San Francisco Book Co.), 1976.
2. Bois, J. S. *Epistemics: The Science-Art of Innovating.* San Francisco: International Society for General Semantics, 1972.
3. Broad, W. J. "Paul Feyerabend: Science and the Anarchist." *Science* 206 (2 November 1979): 534–37.
4. Brown, G. I. *Human Teaching for Human Learning: An Introduction to Confluent Education.* New York: Viking Press, 1971.
5. Burke, J. *Connections.* Waltham, Mass.: Little, Brown, and Co., 1979.
6. Clark, R. *Einstein: The Life and Times.* New York: Thomas Y. Crowell Co., 1971.
7. Downs, R. M., and Stea, D. *Maps in Minds: Reflections on Cognitive Mapping.* New York: Harper and Row, 1977.
8. ES. *ES-SENSE I* and *ES-SENSE II.* Reading, Mass.: Addison-Wesley Publishing Co., 1975.
9. ESTPP. *Gift Garden of Fantaseeds.* Leesburg, Va.: American Geological Institute, 1974.
10. Feyerabend, P. *Against Method.* London: New Left Books, 1975.
11. Fiske, E. B. "Finding Fault with the Testers." *New York Times Magazine,* November 18, 1979, pp. 152–62.
12. Franck, F. *The Zen of Seeing.* New York: Alfred A. Knopf, 1973.
13. Gillie, O. "Did Sir Cyril Burke Fake His Research on Heritability of Intelligence?" *Phi Delta Kappan,* February 1977, pp. 469–71.
14. Goble, F. *The Third Force.* New York: Pocket Books, 1971.
15. Gordon, W. G. G. *Synectics.* New York: Harper and Row, 1961.
16. Gould, P., and White, R. *Mental Maps.* New York: Penguin Books, 1974.
17. Hawkins, D. "Messing About in Science." *Science and Children* 2, no. 5 (1965).
18. Koestler, A. *The Act of Creation.* London: Hutchinson, 1964.
19. Korzybski, A. *Science and Sanity.* Lakeville, Conn.: International Non-Aristotelian Library, 1933.
20. Kozol, J. *The Night Is Dark and I Am Far from Home.* Boston: Houghton Mifflin Co., 1975.
21. Kuhn, T. S. *The Structure of Scientific Revolutions.* Chicago: University of Chicago Press, 1962.
22. Leonard, G. *The Silent Pulse: A Search for the Perfect Rhythm That Exists in Each of Us.* New York: E. P. Dutton and Co., 1978.
23. LeShan, L. *How to Meditate: A Guide to Self-Discovery.* Waltham, Mass.: Little, Brown, and Co., 1974.
24. McCarthy, D. "Parapsychology." In *Science Education/Society: A Guide to Interaction and Influence, 1979 AETS Yearbook,* edited by M. R. Abraham. Columbus, Ohio: ERIC/SMEAC, 1979.

25. Maslow, A. *The Psychology of Science.* New York: Harper and Row, 1966.
26. Michot, P. "Anorthosites et Anorthosites." Academie Royale de Belgique, Classe des Sciences 41, 5E (1955): 275–94.
26a. Osborne, A. E. *Applied Imagination.* New York: Charles Scribner's Sons, 1953.
27. Postman, N. *Teaching as a Conserving Activity.* New York: Delacorte Press, 1979.
28. Prince, G. *The Practice of Creativity.* New York: Harper and Row, 1970.
29. Progoff, I. *Jung, Synchronicity, and Human Destiny.* New York: Julian, 1973.
30. Rockwell, T. "Heresy and Excommunication in American Science." *Association for Humanistic Psychology Newsletter,* November 1979, pp. 11–12.
31. Rogers, C. "The Formative Tendency." *Journal of Humanistic Psychology* 18, no. 1 (1978): 23–26.
32. _____. "Toward a Theory of Creativity." *ETC: A Review of General Semantics* 2 (1954): 249–60.
33. Romey, W. D. "AGI's Boulder Operation." *Geotimes* 17, no. 3 (1972): 24–26.
34. _____. *Confluent Education in Science.* Canton, N.Y.: Ash Lad Press, 1976.
35. _____. *Consciousness and Creativity: Transcending Science, Humanities, and the Arts.* Canton, N.Y.: Ash Lad Press, 1975.
36. _____. "Environmental Studies (ES) Project Brings Openness to Biology Classrooms." *American Biology Teacher* 34 (1972): 322–28.
37. _____. "The Essence of Life." In *Science Education—Society: A Guide to Interaction and Influence, 1979 AETS Yearbook,* edited by M. R. Abraham. Columbus, Ohio: ERIC/SMEAC, 1979. (Contains extensive annotated bibliography.)
38. _____. *Inquiry Techniques for Teaching Science.* Englewood Cliffs, N.J.: Prentice-Hall, 1968.
39. _____. "Introductory Geology from the Newspapers." *Journal of Geological Education* 25 (1977): 111–14.
40. _____. "Zen in the Art of Science Teaching." *School Science and Mathematics* 78 (1978): 115–23.
41. Rosenthal, R., and Jacobson, L. *Pygmalion in the Classroom: Teacher Expectation and Pupil's Intellectual Development.* New York: Holt, Rinehart, and Winston, 1968.
42. Russell, B. "Education and Discipline." In *In Praise of Idleness and Other Essays,* pp. 202–10. New York: Simon and Schuster, 1972.
43. _____. "Modern Inhomogeneity." In *In Praise of Idleness and Other Essays,* pp. 188–98. New York: Simon and Schuster, 1972.
44. St. James-Roberts, I. "Cheating in Science." *New Scientist,* November 25, 1976, pp. 466–69.
45. Schwab, J. *The Biology Teacher's Handbook.* New York: John Wiley and Sons, 1963.
46. Stevens, J. *Awareness: Exploring, Experimenting, Experiencing.* Moab, Utah: Real People Press, 1971.
47. Thompson, J., and others. "Opening Learning Environments: The Ultimate Individualization." In *Individualized Science–Like It Is,* edited by H. Triezenberg, pp. 44–60, 93–99. Washington, D.C.: National Science Teachers Assn., 1972. (Contains extensive annotated bibliography.)

48. Tuttle, F. B., Jr. *Gifted and Talented Students.* Rev. ed. Washington, D.C.: National Education Association, 1985.
49. _____. "Providing for the Intellectually Gifted." *SLATE.* Urbana, Ill.: National Council of Teachers of English, November 1979.
50. _____, and Becker, L. A. *Characteristics and Identification of Gifted and Talented Students.* Washington, D.C.: National Education Association, 1980.
51. _____. *Program Design and Development for Gifted and Talented Students.* Washington, D.C.: National Education Association, 1980.
52. Vernon, P. E., ed. *Creativity.* Harmondsworth, Middlesex: Penguin, 1970.
53. Wade, N. "IQ and Heredity: Suspicion of Fraud Beclouds Classic Experiment." *Science* 194 (1976): 916–19.
54. Watson J. D. *The Double Helix: Being a Personal Account of the Discovery of the Structure of DNA.* New York: Atheneum Publishers, 1968.

# SELECTED RESOURCES

Barrington, B. L. "Curriculum-Based Programs for the Gifted." *Education Digest* 52 (January 1987): 48–51.

Bellamy, M. L. "What's Your Theory? Working on Problems with Unknown Solutions." *Science Teacher* 50 (February 1983): 34–36.

Blurton, C. "Individualized Science Packets for Gifted Students." *School Science and Mathematics* 83 (April 1983): 326–32.

Blurton, C. "Science Talent: The Elusive Gift." *School Science and Mathematics* 83: (December 1983): 654–64.

Blurton, C., and Staley, F. "Science and the Gifted Young Child." *Early Years* 12 (April 1982): 28.

Cohen, H. G. "Science and the Young Gifted." *Clearing House* 56 (April 1983): 374–77.

Day, F. "Adding Drama to Science." *Science and Children* 19 (February 1982): 15.

Finley, F. N., and others. "Teachers' Perceptions of Important and Difficult Science Content." *Science Education* 66 (July 1982): 531–38.

Hassard, J. "Opening the Mind's Eye to Science." *Science and Children* 19 (April 1982): 30–32.

Johnson, T. "Classroom Drama." *School and Community* 69 (May 1983): 34.

Lowery, J. "Developing Creativity in Gifted Children." *Gifted Child Quarterly* 26 (Summer 1982): 133–39.

Padilla, M. J., and others. "The Relationship Between Science Process Skill and Formal Thinking Ability." *Journal of Research in Science Teaching* 20 (March 1983): 239–46.

Penick, J. E. "What Research Says: Encouraging Creativity." *Science and Children* 20 (February 1983): 30–33.

Scobie, J., and Nash, W. R. "A Survey of Highly Successful Space Scientists Concerning Education for Gifted and Talented Students." *Gifted Child Quarterly* 27 (Fall 1983): 147–51.

Smith, R. L. "Computers and the Gifted." *Journal of Computers in Mathematics and Science Teaching* 5 (Winter 1985–86): 70–71.